A Student Guide to
SPSS

Second Edition

Carrie Cuttler

Concordia University

Kendall Hunt
publishing company

Cover image © Shutterstock, Inc.

publishing company

www.kendallhunt.com
Send all inquiries to:
4050 Westmark Drive
Dubuque, IA 52004-1840

Printed in the United States of America
10 9 8 7 6 5 4 3 2 1

CONTENTS

PREFACE

This second edition of *A Student Guide to SPSS* is packaged with an access card for a free download of IBM® SPSS® Statistics Base Integrated Student Edition Version 22. SPSS stands for Statistical Package for the Social Sciences. As its name implies, the software was developed for use in the social sciences, and it is currently one of the most commonly used statistical software packages in psychology, business, education, and market research. The popularity of SPSS is likely a result of its intuitive drop-down menus that allow for quick and easy computation of some of the most commonly used statistics.

This guide will provide you with easy to follow, step-by-step instructions on how to conduct some of the most frequently used functions in SPSS, as well as how to interpret and report the results. While this guide was originally developed for use in introductory statistics courses, numerous updates have been made to the second edition in order to expand the guide's suitability for more advanced courses in statistics (e.g., advanced undergraduate statistics or introductory graduate level statistics courses). Independent learners with a rudimentary knowledge of statistics will also find this a useful introduction to computing, interpreting, and reporting some of the most commonly used statistics in the social sciences.

New to the Second Edition

This second edition of *A Student Guide to SPSS* has been updated to be compatible with version 22 of SPSS. It includes a free download of IBM® SPSS® Statistics Base Integrated Student Edition Version 22 and features all new screenshots. Throughout the guide, elaborations on the meaning and interpretation of the various statistics, and demonstrations of more advanced statistical analyses have been added. The statistical notation and abbreviations approved for use in the sixth edition of the *Publication Manual of the American Psychological Association* (informally known as the APA style guide) have been introduced, and guidelines on reporting results have also been updated to be consistent with APA style. More specific changes to the various chapters are listed below.

- Chapter 1 now contains a description of the split file function in SPSS. In addition, the APA guidelines on reporting decimal remainders are now reviewed.

- Chapter 2 has been updated to include information on how to compute and interpret skewness and kurtosis statistics.

- Chapter 3 now contains information on how to determine whether a correlation coefficient is statistically significant, and how to compute and interpret one-tailed correlation analyses.

- Chapter 4 has been expanded to include information on how to determine whether a regression model and the individual predictors contained in the model are statistically significant. This chapter also includes new information on the interpretation of standardized regression coefficients (i.e., beta weights), as well as descriptions of how to compute and interpret partial, semi-partial, and squared semi-partial correlation coefficients.

- Chapter 5 is an entirely new chapter on advanced regression. This chapter features step-by-step instructions on computing, interpreting and reporting the results of hierarchical and stepwise regression.

- Chapter 6 (which previously contained information on single-sample *t* tests) and Chapter 7 (which previously contained information on paired-samples and independent-samples *t* tests) have been consolidated into one chapter (Chapter 7) on all three types of *t* tests. Chapter 7 also now includes information on how to compute, interpret, and report Cohen's *d* values for single-sample, paired-samples, and independent-samples *t* tests. It also includes information on conducting one-tailed *t* tests.

- Chapter 8 is an entirely new chapter on one-way between groups ANOVA. It includes demonstrations of computing and interpreting the effect size indicator eta-squared, as well as performing and reporting the results of post hoc tests.
- Chapter 9 is another new chapter on between groups factorial ANOVA. It includes demonstrations of a 2×2 ANOVA with follow up simple main effects analyses, as well as a 2×3 ANOVA with follow up simple main effects analyses and post hoc comparisons.

A LIST OF STATISTICAL SYMBOLS AND ABBREVIATIONS USED IN THIS GUIDE

Symbol/Abbreviation	Meaning
a	Intercept
ANOVA	Analysis of Variance
b	Slope (unstandardized)
CI	Confidence interval
d	Cohen's d
df	Degrees of freedom
HSD	Honestly significant difference (Tukey's)
F	F statistic
M or $\overline{X}$	Mean (sample)
M_D	Mean of the difference scores
Mdn	Median
N	Sample size (total sample)
n	Sample size (subsample)
p	p value
r	Pearson correlation
R	Multiple correlation
r^2	Coefficient of determination
R^2	Coefficient of multiple determination
$r_{ab.c}$	Partial correlation
$r^2_{ab.c}$	Squared partial correlation
$r_{a(b.c)}$	Semi-partial correlation
$r^2_{a(b.c)}$	Squared semi-partial correlation
r_{pb}	Point biserial correlation
r_s	Spearman rank order correlation
s or SD	Standard deviation (sample)
SD_D	Standard deviation of the difference scores
s^2	Variance (sample)
SE	Standard error
SEE	Standard error of estimate
SEM	Standard error of the mean
SS_B	Sum of squares between
SS_W	Sum of squares within
SS_{Total}	Total sum of squares
t	t statistic
t_{crit}	Critical value for t statistic
X	Predictor variable
Y' or $\hat{Y}$	Criterion variable
z	z score (standardized score)

Greek Characters	Meaning
β	Slope (standardized)
Δ	Change
η^2	Eta-squared
η_p^2	Partial eta-squared
μ	Mean (population)
μ_{lower}	Lower limit of a confidence interval
μ_{upper}	Upper limit of a confidence interval
σ	Standard deviation (population)
σ^2	Variance (population)
ϕ	Phi coefficient

ABOUT THE AUTHOR

Carrie Cuttler is currently working as an Assistant Professor in a limited-term appointment at Concordia University in Montreal, Quebec. She received her PhD in cognitive psychology from the University of British Columbia (UBC) in 2008. She subsequently completed a postdoctoral fellowship in the Department of Psychiatry in 2011.

Her diverse research interests are united by a focus on individual differences in prospective memory, which is the ability to remember to execute delayed intentions (e.g., remembering to hand in your statistics assignment). Specifically, her research targets populations (e.g., individuals with checking compulsions, chronic cannabis users, pregnant women), traits (e.g., Big Five personality traits, impulsiveness, perfectionism), and states (e.g., anxiety, depression, stress) likely to be associated with prospective memory problems.

Carrie has been teaching undergraduate courses in research methods and statistics since 2007. This guide was originally developed as part of a project to create a practical laboratory component for UBC's introductory statistics course, but has since taken on a life of its own. Carrie also wrote *Research Methods in Psychology: Student Lab Guide* as part of a sister project to create a laboratory component for UBC's introductory research methods course. Carrie strives to connect her roles as a researcher and educator by applying her knowledge of human learning and memory to the classroom. Her interest in developing laboratory components for university level courses stems from her awareness of the importance of the practical application of course material to promote deeper and longer-term learning.

Carrie would love to hear from you. If you have any comments or concerns about this guide please email Carrie at: astudentguidetospss@gmail.com.

An Introduction to SPSS

1

Learning Objectives

In this chapter, you will learn how to create variables, define the properties of those variables, and enter data. You will also learn some handy tools and tricks for manipulating and transforming data.

GETTING STARTED

Installing SPSS

The enrollment code that is packaged with this guide entitles you to a free download of IBM® SPSS® Statistics Base Integrated Student Edition Version 22. The Mac version of this software is compatible with Mac OS X Lion 10.7 and Mountain Lion 10.8[1]; and the Windows version is compatible with Microsoft® Windows XP, Windows Vista®, Windows® 7, and Windows® 8. The software expires after 12 months, can only be installed once, and is limited to 50 variables (columns) and 1500 cases (rows). The student version of this software cannot be used to conduct certain advanced analyses (e.g., repeated measures ANOVA) that full versions of the software are capable of performing, and syntax codes cannot be written using this version of the software. Before attempting to install SPSS you must first be logged onto your computer with administrator privileges. To install the software, simply follow the instructions written on the inside of the front cover of this guide.

Opening SPSS

You can open the SPSS program by clicking on the program icon 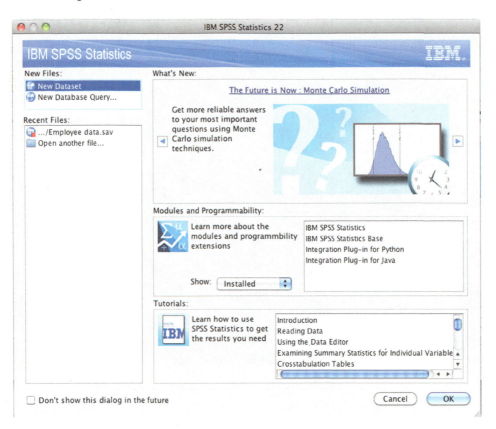 or the filename 'SPSS Statistics.' If you used the default installation, the program will be in your applications (Mac) or program files (PC) in a folder labeled 'IBM.' Specifically, you will need to go to IBM→SPSS→Statistics→22→SPSSStatistics. Note that it is a large program, so it can take a couple of minutes for it to open. When you first open the program, a window like the one shown below will appear. It provides you with several options, including options to run tutorials, open recently opened files, or open a new dataset. For now, select the option in the top left corner to open a **New Dataset** and then click **OK**.

---

[1] Although I'm currently using it with Mac OS X Snow Leopard 10.6.8 and it is working just fine.

Note that if you want to open an existing dataset that has already been saved on your computer, you can simply click on the data file in your computer menu.[2] Alternatively, you can click on the option to 'Open another file' in the window shown on the preceding page and then locate and open the data file.

DATA AND VARIABLE VIEW WINDOWS

Data View

Once SPSS is open, the Data View window will appear. As its name implies, this is where you enter the data or where the data will appear if you are opening an existing dataset. Along the top of the screen you will see a toolbar with different options (e.g., File, Edit, View, Data, Transform, Analyze, etc.), and another toolbar with a series of different icons.

For now, let's focus on the matrix of cells occupying most of the screen. Along the top of this matrix, you will see a row of cells each labeled 'var' to represent variable. Along the left side of this matrix, you will see a column of numbered cells. The columns in Data View are used to designate the different variables. The rows in Data View are used to designate the different participants (or cases) assessed. Thus, any given cell represents a specific participant's score on a specific variable.

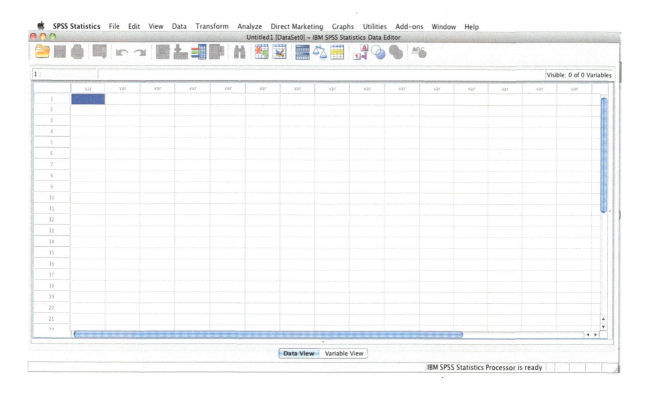

You can highlight cells in the matrix by clicking on them. Pressing enter will move you down one cell. Take a second now to practice moving around the cells using your trackpad, mouse, and/or arrow keys.

---

[2] Troubleshooting tip: If an existing SPSS data file won't open when you double click on its filename in your computer menu, then try to first open a blank SPSS spreadsheet using the 'New Dataset' option. Once a blank SPSS spreadsheet is open, go to File→Open→Data (or click the orange folder in the icon toolbar), locate the file on your computer, highlight the file, and then click Open.

At the bottom of the screen, you will see a tab labeled 'Data View' and a tab labeled 'Variable View.' As shown in the image displayed on the preceding page, the tab labeled Data View is currently highlighted because you are looking at the Data View window. Click the **Variable View** tab to open the Variable View window.

Variable View

The Variable View window is used to enter in the names of the variables and to define the properties of those variables (or in the case of an existing dataset, it can be used to learn about the properties of the different variables). As described fully below, the variable names need to be entered in the first column and their properties can be defined using the remaining columns. Once the variable names are entered in the first column of the Variable View window, they will appear in the top row of the Data View window in place of 'var.' Note that the display in Variable View is transposed from the one in Data View; the Variable View window shows the variable names in the first column, while the Data View window shows the variable names in the top row.

The column labeled 'Name' is where you need to enter the names of your variables. The column labeled 'Type' is where you can indicate the type of data you will enter (e.g., numeric, currency, string). The column labeled 'Width' is where you can set the maximum number of characters that the values of the variables can contain. The column labeled 'Decimals' is where you can set the number of decimal places you want displayed for the values of the variables. The column labeled 'Label' is where you can enter text descriptions of the variables. The column labeled 'Values' is where you can define any codes you are using to identify the values of the variables. The column labeled 'Missing' permits you to enter any special code (e.g., −99) that you used to identify missing values. The column labeled 'Align' can be used to indicate how you would like the data in the Data View window to be aligned (right, left, or center justified). The column labeled 'Measure' is where you can specify the scales of measure used to measure the variables. Note that SPSS uses the term 'Scale' to refer to both interval and ratio scales. Finally, the column labeled 'Role' can be used to define whether the variable is a predictor or outcome (criterion) variable; it is a setting that is not generally used, and as such, we will not consider it further.

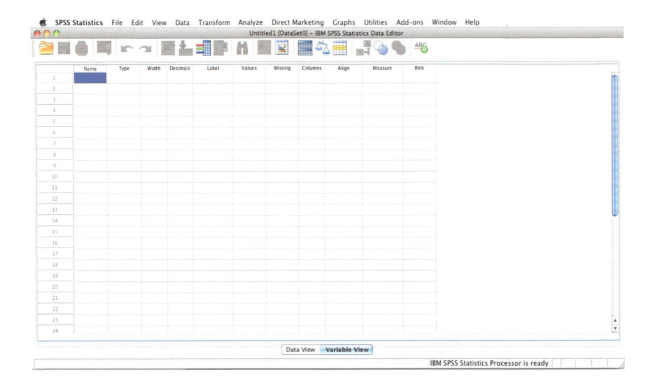

Creating Variables (in Variable View)

Let's get started by creating some variables. We will begin by creating a participant ID variable. While you are in the Variable View window, type **ID** in the first cell in the **Name** column (the cell that is highlighted blue in the image displayed above). Variable names must be one word; spaces and special characters like / and - are not allowed. An error message stating that the variable name contains an illegal character will appear if you attempt to use one of these forbidden characters. The ID codes we will use will be numeric and they will be under 8 digits long so you can leave the Type as the default Numeric and the Width at the default of 8 characters. Since we do not want our ID codes to contain a decimal remainder, we will need to change the default of 2 to 0. To do this, you will need to move over to the **Decimals** column, then click on the **down arrow button** that appears when you click on this cell until a value of **0** appears. Our variable name is quite informative, but this is not always the case, so we should practice entering a label for the variable. Click on the first cell in the **Label** column and type **Participant ID Codes**. Finally, set the scale of measure to nominal by moving over to the **Measure** column and clicking on the option for **Nominal**. Now our first variable has been created and its properties have been defined!

Next, let's create a weight variable to represent the weight of each participant in pounds. Type the word **Weight** into the second cell in the **Name** column (immediately under ID). The weights we will enter will be in numeric format, so leave the Type as Numeric. Once again, by default the value in the Width column is set to 8 and the value in the Decimals column is set to 2. Since the weights we will enter will be under 8 digits long, you do not need to adjust these settings. To provide a meaningful description of the variable, you should type **Weight in Pounds** in the second cell in the **Label** column. Finally, since weight is measured on a ratio scale, you should set the scale of measure to scale (the term SPSS uses for interval and ratio scales) by clicking on the relevant cell in the **Measure** column and then clicking on the word **Scale**.

Let's also create a gender variable. Type the word **Gender** into the third cell in the **Name** column (immediately under Weight). We will use numeric codes to represent the different genders, so you should leave Type as Numeric. We will change the width of the cells to 1, but before we can do so we need to change the decimals to 0. Start by clicking on the relevant cell in the **Decimals** column, and then click on the **down arrow button** until a value of **0** is displayed. Now we can change the width of the cells to 1 by highlighting the relevant cell in the **Width** column and clicking the **down arrow button** until it displays a value of **1**. To label the variable, type **Gender** in the third cell in the **Label** column.

SPSS will not permit analyses on string variables (variables containing text), so if we want to perform any analyses with the gender variable, we will need to use numeric codes to label the genders. We will use the number 1 to designate the male gender, and the number 2 to designate the female gender. To assign these numeric codes, move over to the adjacent cell in the **Values** column and click the little **grey box with three dots** that appears when the cell is highlighted. A 'Value Labels' dialog window like the one shown on the following page will now appear. Enter a **1** in the **Value** box and **Male** in the **Label** box. Click **Add**. You just told SPSS that the gender code 1 means male. Now enter **2** in the **Value** box and **Female** in the **Label** box. Click **Add**. Now SPSS knows that the gender code 2 means female. Click **OK** to close the dialog window. Finally, change the scale of measure to nominal by clicking on the relevant cell in the **Measures** column and then clicking on the word **Nominal**.

Let's create one last variable. Type the word **Ethnicity** into the fourth cell in the **Name** column (immediately under Gender). By default SPSS only permits numbers to be entered in the Data View spreadsheet. To change this default setting, you need to click on the relevant cell in the **Type** column and then click the little **grey box with three dots** that appears when the cell is highlighted. A dialog window will now appear. Click on the **String** option, and then click **OK** to close the dialog window. Now we will be able to enter ethnicities using text rather than numeric codes. Changing the type to string automatically changes the decimals to 0, the alignment to left, and the measure to nominal (since any data that contain letters must be nominal/categorical in nature). Change the width of the cells to 9 by clicking on the relevant cell under the **Width** column and then clicking the **up arrow button** until it displays a value of **9**. To label the variable, type **Ethnicity** in the third cell under the **Label** column. Since we will be using text rather than codes to enter participants' ethnicities, there is no need to define the values of the variables.

Your Variable View window should now look like the one shown below.

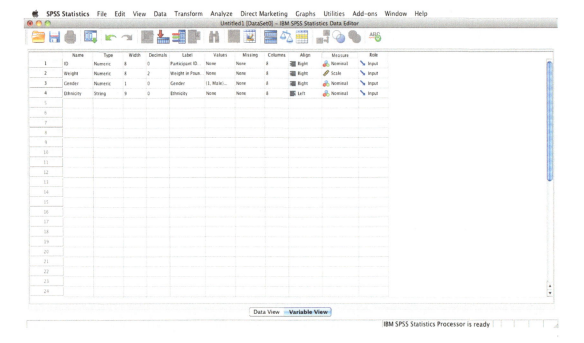

Now return to the Data View window by clicking on the **Data View** tab at the bottom of the screen. Voilà! As highlighted in the image shown below, the variable names we created now appear in the top row of the Data View window.

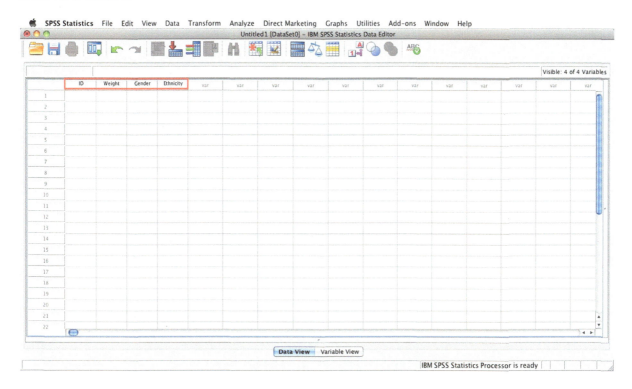

Entering Data (in Data View)

Recall that in the Data View window, each column represents a variable and each row represents a single participant's scores on each of the variables. As such, we will need to enter the participant ID codes into the first column, the participants' weights into the second column, the participants' gender codes into the third column, and the participants' ethnicities into the fourth column. Try entering the following data into the spreadsheet by clicking on the relevant cell and typing in the information. Do not enter the variable names shown in the top row, as they should already appear in the top row of your Data View spreadsheet.

ID	Weight	Gender	Ethnicity
1	105	2	Asian
2	165	1	Caucasian
3	135	2	Caucasian
4	185	1	Caucasian
5	120	2	Asian

Once the data are entered your screen should look like the one shown on the next page. It should now be clear that each row represents a different participant, and each column represents a different variable. Thus, each cell represents one participant's score on one of the variables. The first participant's ID code is 1. That participant weighs 105 lbs, is female (which we coded 2), and is Asian. The second participant's ID code is 2. That participant weighs 165 lbs, is male (which we coded 1), and is Caucasian. The third participant's ID code is 3. She is a Caucasian female weighing 135 lbs. The fourth participant's ID code is 4. He is a Caucasian male weighing 185 lbs. The last participant's ID code is 5. She is an Asian female who weighs 120 lbs.

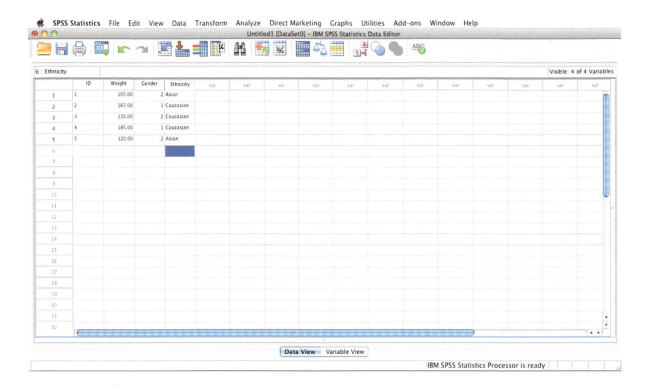

SAVING THE SPSS DATA FILE

Now that we've done all of this work, we want to make sure we don't lose it (we will use this dataset in the remainder of this chapter and again in Chapter 2). You can save the file either by pressing the **disk icon** in the icon toolbar, or by clicking **File→Save As**. When prompted, type in the filename **Practice Data**, find a good location on your computer to save it, and press **Save**. An output window will appear displaying your command to save the data file in computer code. You can simply close that window for now. A prompt to save the output window will appear when you try to close it. Since we only want to save the data file (and not the code generated to save the data), you can click **No**.

> Congratulations, you now know how to open SPSS, create variable names, define the properties of the variables, enter data, and save your data. You're off to a great start!

HANDY TOOLS AND TRICKS

Viewing Value Labels

Typically, in order to decode variable codes (e.g., gender codes), you need to refer to the Values column in the Variable View window. However, a shortcut is available allowing you to quickly view variable codes in the Data View window. This is useful if you forget what your codes represent or if you are using a data file that someone else prepared and you do not know what the various codes represent. To decode variables that are coded, press the **Value Labels icon** in the icon toolbar (highlighted in the image shown below). Once you click this icon, you will see that all of the gender codes (the 1s and 2s) change to the labels of the codes we previously entered in the Value Labels dialog window (Male and Female). To recode the variable, simply press the Value Labels icon again.

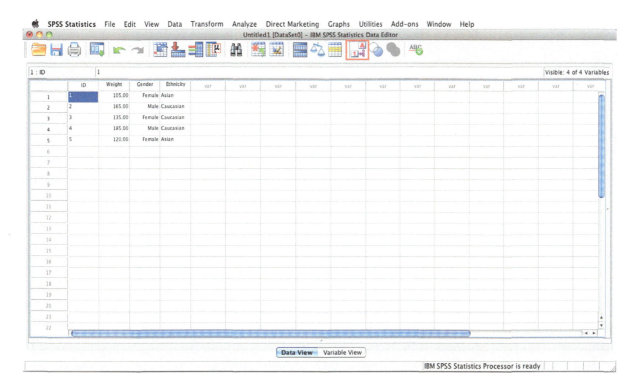

Sorting Data
(Data→Sort Cases)

It is often useful to sort your data according to the values of a variable (i.e., to reorder scores from lowest to highest or highest to lowest). For instance, if we wanted to find the most extreme scores on a variable, we could sort them and then look at the scores at the very top and very bottom of the column. Let's practice using this function by sorting the data according to weight. Go to the upper toolbar and click **Data→Sort Cases**.

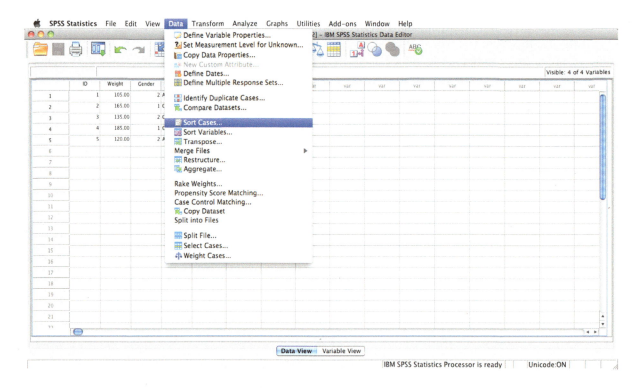

A 'Sort Cases' dialog window like the one displayed below will now appear. Highlight the **Weight** variable shown on the left side of the dialog window by clicking on it, and then press the **blue arrow** to move it over to the box on the right labeled 'Sort by.' Notice you can choose to sort from lowest to highest (ascending) or from highest to lowest (descending). Leave it at the default, which is to sort in ascending order, and press **OK** to close the window.

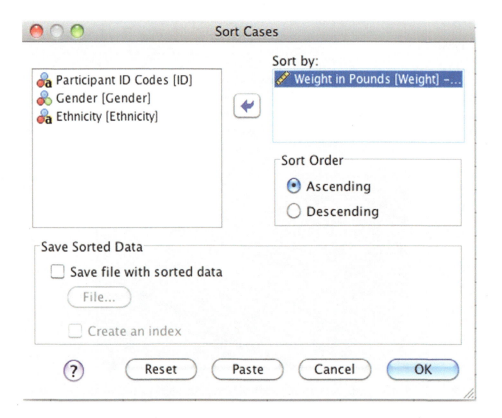

An output window will appear displaying the command to sort the data. You can simply close that window. Once again, there is no need to save the output window because it only contains the code for the command, so select **No** when prompted to save the output.

Notice that the data are now organized such that the lightest participant's data (ID 1) are in the first row and the heaviest participant's data (ID 4) are in the last row. You should practice using this feature by re-sorting the data so they are once again organized according to ID code.

Selecting Cases
(Data→Select Cases)

Another handy tool is selecting cases. This tool allows you to perform calculations and analyses using only a specific subset of the participants. For example, if you wanted to know the average weight of only the male participants, you would first need to select the male participants' data before computing the mean weight (computation of the mean is described in Chapter 2). Let's practice by selecting only the male participants' data. Go to the upper toolbar and click **Data→Select Cases**. Alternatively, you can simply click on the **Select Cases icon** in the icon toolbar (highlighted in the image shown below).

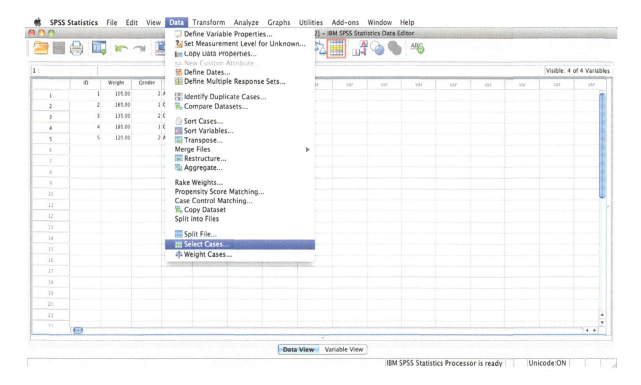

A 'Select Cases' dialog window like the one displayed on the next page will now appear. The default is to have all cases selected (you can think of cases as participants). To select only a subset of the cases, select **If condition is satisfied** and press the **If button**.

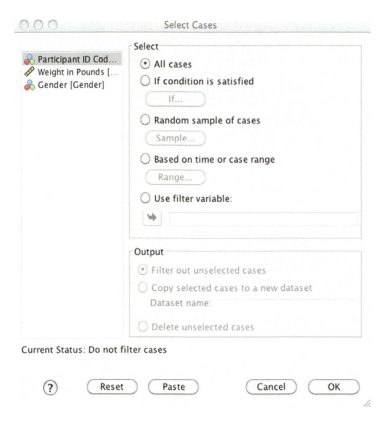

A 'Select Cases: If' dialog window like the one shown below will now open. Highlight the **Gender** variable located on the left side of the dialog window by clicking on it, then move it over to the box on the upper right side of the window by clicking on the **blue arrow**. Next to Gender enter **= 1** using the onscreen keypad. The box should now read Gender = 1, as shown below. Press **Continue** to close this dialog window. Press **OK** to close the Select Cases dialog window. Close the output window that appears without saving.

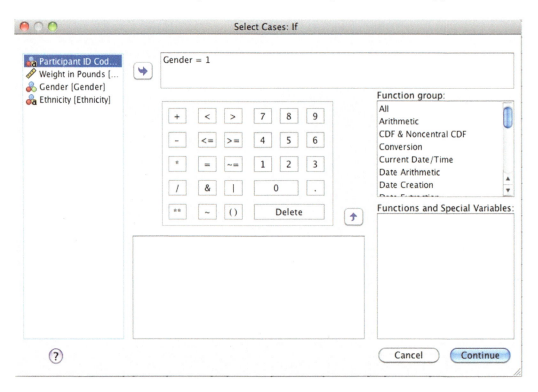

As shown in the image displayed below, lines will now appear through the cells next to the three female participants. The lines indicate that these participants will be dismissed or excluded from future analyses. You should also notice a new column labeled 'filter_$' has been created (highlighted blue in the image shown below). The 0s in that column indicate the participants who will be excluded from subsequent analyses because they don't meet the criteria we specified (females), and the 1s indicate the participants who will be included in subsequent analyses because they do meet the criteria we specified (males). Once again, you can also reveal these codes by clicking the Value Labels icon in the icon toolbar. If you use this option, the words 'Selected' and 'Not Selected' will appear rather than 1s and 0s.

Note that if you wanted to perform subsequent analyses on only the selected cases, you would still use the variable of interest when performing the analysis. For instance, if you wanted to calculate the mean weight of only the male participants, you would set the filter to select only males and then you would calculate the mean using the weight variable. Never attempt to conduct an analysis using the 'filter_$' variable as a variable. The column is only created to indicate which participants have been filtered out; it is not a real variable and cannot be analyzed in any meaningful way.

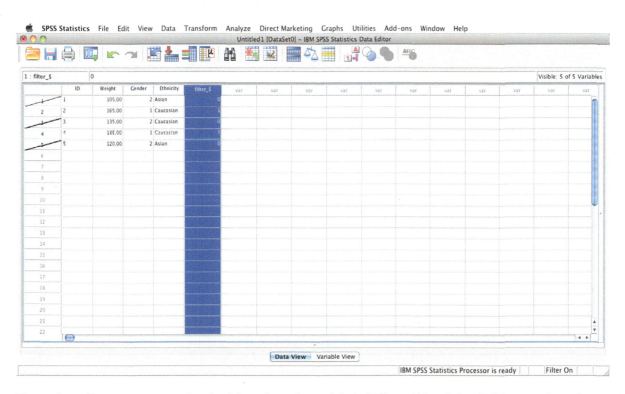

To reselect all cases, you can simply delete the column labeled 'filter_$' by **right clicking** on the **column header** (the filter_$ label) and clicking **Cut**. Alternatively, you can click on the **Select Cases icon** in the icon toolbar, or go to **Data→Select Cases**, and then using the 'Select Cases' dialog window, click **All Cases**, then press **OK**. If you use this alternative option, you will notice that the lines through the cells disappear, but the 'filter_$' column remains (although it is now inactive).

Next, let's try selecting participants with weights less than or equal to 165. Click on the **Select Cases icon** in the icon toolbar or go to **Data→Select Cases**. Select **If condition is satisfied** and press the **If button**. Delete any information in the upper white box (Gender = 1 will still be in there), then highlight the **Weight** variable on the left side of the dialog window by clicking on it and move it into upper white box by clicking on the **blue arrow**. Next to Weight enter **<= 165** using the onscreen keypad (note: <= designates less than or equal to, while => designates greater than or equal to). Your screen should now look like the one displayed on the next page. Click **Continue** and then click **OK** to close the dialog windows. Close the output window without saving the output.

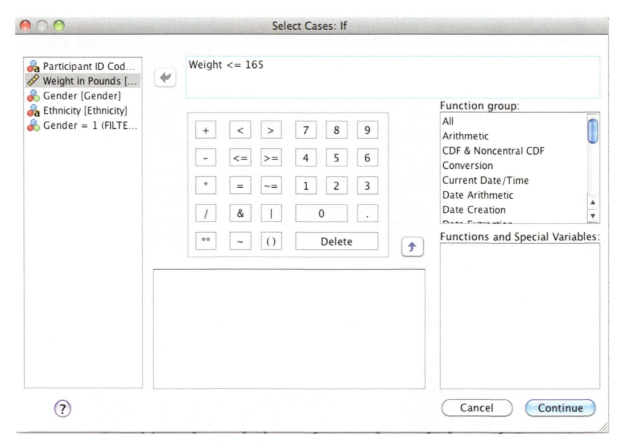

Now the participant with a weight over 165 will not be entered into any future analyses and your Data View window will show a line through the cell next to his participant ID. Try reselecting all of the cases on your own so that this participant will not be excluded from our subsequent computations and analyses.

Splitting Data into Groups
(Data→Split File)

The split file option is useful when you want to perform separate analyses on subsets of the participants. This tool allows you to split your file into subgroups of participants so that they can be considered separately. For instance, if we wanted to know the average weight of males and the average weight of females separately, we could use this option to split our file by gender before computing the mean weight. Let's practice using this tool by splitting the data file by gender. Go to **Data→Split File**.

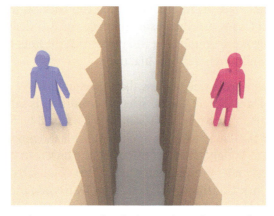

© sibgat, 2013. Used under license from Shutterstock, Inc.

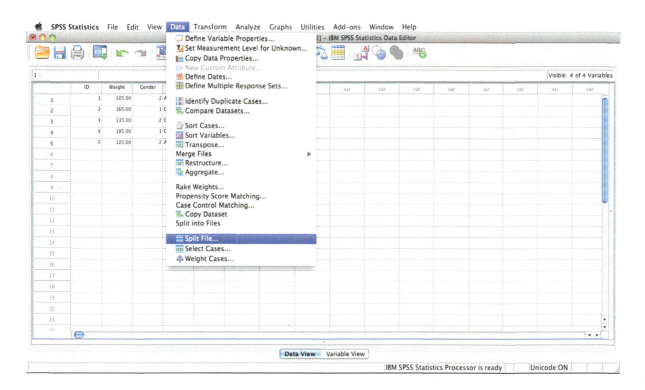

A 'Split File' dialog window like the one shown below will appear. Click **Organize output by groups** and then highlight the variable name **Gender** and move it over to the **Groups Based on box** by clicking on the **blue arrow**. Click **OK**. Close the output window without saving the output.

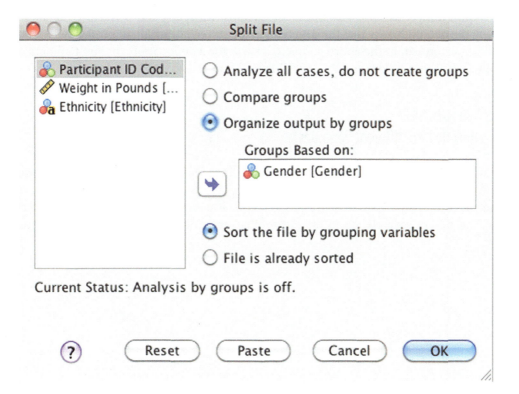

As shown in the display below, the data are now sorted by gender. In addition to the data being reorganized, any subsequent analyses will now be performed separately for males and females, and the results of these analyses will be presented for each of these groups separately.

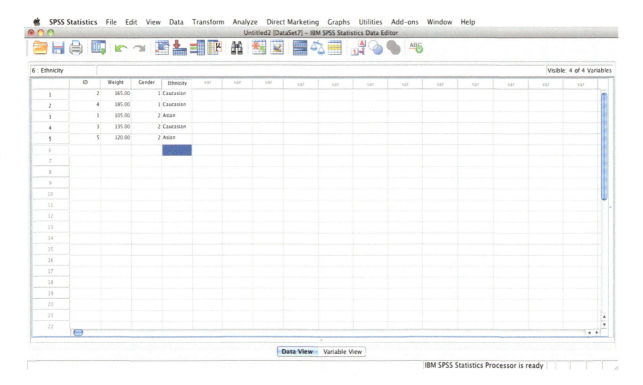

Since we do not want our data file to remain split, we should deactivate this feature by going to **Data→Split File** and selecting **Analyze all cases, do not create groups**. Click **OK** and then close the output window without saving the output. Notice that the data remain organized according to gender, but any subsequent analyses will be performed on the entire sample. This is a good opportunity for you to practice using the sort cases function to reorganize the data according to ID codes.

Recoding Variables
(Transform→Recode into Different Variables)

Sometimes we need to recode a variable. For instance, since we entered the participants' ethnicities as a string (text) variable, SPSS will not allow us to perform any analyses with the variable. In order to perform analyses on this variable, we will need to recode it using numeric codes. To recode an existing variable into a different variable, go to the top toolbar and click **Transform→Recode into Different Variables**.[3]

---

[3] We could also use the option to 'Recode into Same Variables,' but this would overwrite our original variable.

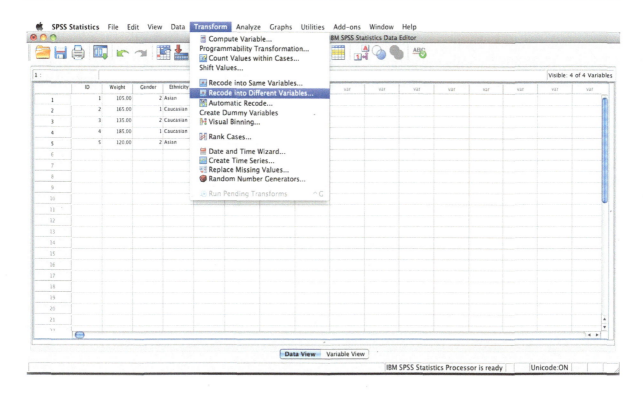

A dialog window labeled 'Recode into Different Variables' like the one shown below will now appear. Move the variable you want to recode—in this case **Ethnicity**—into the box labeled **String Variable→ Output Variable** by highlighting the variable name in the box on the left and clicking on the **blue arrow**. Enter the name of the new variable you want to create—in this case **EthnicityCode**—in the box labeled **Name**. Click on the **Change button** (highlighted blue in the image shown below).

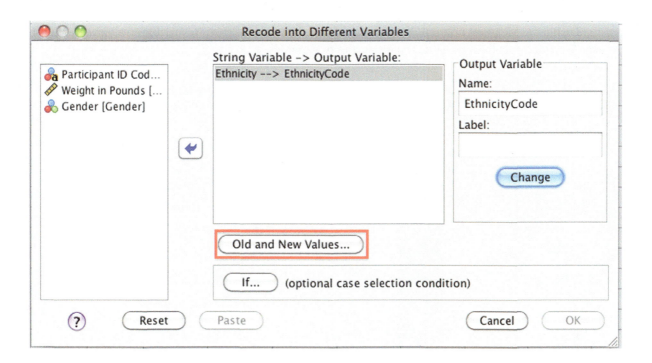

Next, click the button labeled **Old and New Values** (highlighted red in the image shown on the preceding page). A new dialog window like the one shown below will now appear. We will recode Caucasian as 1 and Asian as 2. To do this you will need to enter **Caucasian** into the **Old Value: Value box** and enter a **1** into the **New Value: Value box**. Click **Add**. Next, enter **Asian** into the **Old Value: Value box** and enter a **2** into the **New Value: Value box**. Click **Add**. Finally, click **Continue** and then **OK** to close the dialog windows. Close the output window that will appear without saving it.

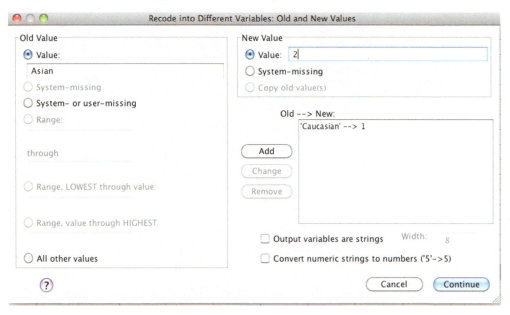

As highlighted in the image shown below, a new variable labeled 'EthnicityCode' will now appear in the Data View window. Caucasian participants are labeled with a 1 and Asian participants are labeled with a 2. You should now go to the Variable View window and define the properties of this new coded variable. You should change the number of Decimals to 0 and the Width to 1. Label the variable 'Ethnicity Coded,' define the values of this new coded variable using the Values column, and define the scale of measure as nominal using the Measures column. You should refer to the section Creating Variables (in Variable View) on page 5 if you forget how to define the properties of variables.

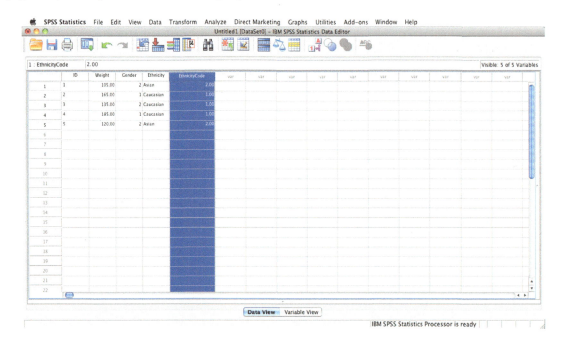

Computing New Variables
(Transform→Compute Variable)

Sometimes we need to compute a new variable by transforming an existing variable. For instance, let's say we wanted to compute all of the participants' weights in kilograms. To compute a new variable using an existing variable, you need to go to the top toolbar and click **Transform→Compute Variable**.

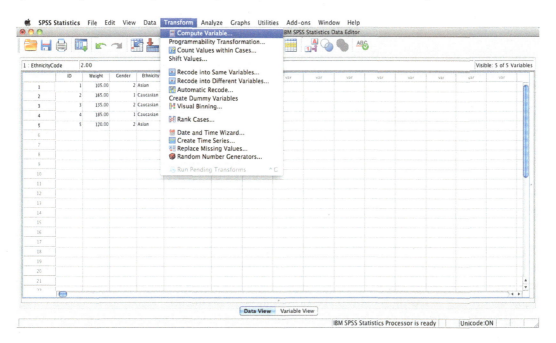

A 'Compute Variable' dialog window like the one shown below will now appear. Type the name of the new variable you want to create—**Kg**—in the **Target Variable box**. Highlight the **Weight** variable on the left side of the dialog window and move it into the **Numeric Expression box** using the **blue arrow**. One pound equals 0.45 kilograms, so to convert all of the weights from pounds to kilograms you will need to multiply each weight by 0.45. To do this, enter * **0.45** next to the variable name Weight (the * indicates multiplication; a / would indicate division). Your screen should now look like the one shown below. Press **OK** to close the dialog window. Close the output window without saving.

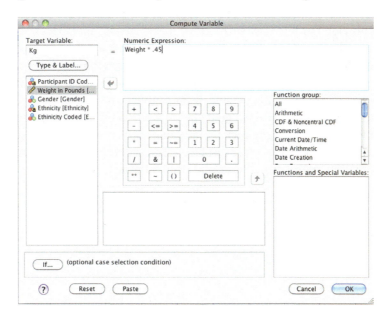

A new variable labeled 'Kg' will now appear in your Data View window, with all of the participants' weights in kilograms. You can perform just about any mathematical transformation to your data using this handy tool!

Try On Your Own

I encourage you to explore the many handy tools and tricks available in the icon toolbar and under the Data and Transform menus in the upper toolbar. You may wish to play around with these functions on your own or by following some of the tutorials, which can be found using the Help menu in the top toolbar.

REPORTING DECIMAL REMAINDERS AND ROUNDING

Reporting Decimal Remainders

The sixth edition of the *Publication Manual of the American Psychological Association* (the APA style guide) advises us to report our results to two decimal places. There are only a few exceptions to this general rule. First, discrete variables, which are variables that cannot have a decimal remainder (e.g., number of participants, degrees of freedom), should be presented with no decimal remainder. Also, p values (and other values) between .001 and .009 should be reported to three decimal places, while p values lower than .001 should be reported as $p < .001$.

Statistics that cannot take on values greater than 1 (e.g., correlation coefficients, p values), should be reported without a leading 0 before the decimal place (e.g., $p = .03$). Statistics and other values that can be greater than 1 (e.g., standard deviations, variance), should be reported with a leading 0 before the decimal place when the values are lower than 1 (e.g., $s^2 = 0.36$).

Rounding

The results of analyses provided by SPSS will be rounded to various decimal places; sometimes no decimal remainders are shown, while other times five or more decimal places are provided. As such, some rounding is typically required. If you want to round to two decimal places and the value in the third decimal place is lower than 5, you should round down, but if the value in the third decimal place is equal to or greater than 5, you should round up. When a table in the output window shows a value rounded to three decimal places and the number in the third decimal place is a 5, you will need to determine if the value has been rounded up or down. To do so, simply click on the value in the output window several times until the entire string of digits after the decimal place is displayed. You will then be able to make a decision about whether you need to round the number in the second decimal place up or down.

Descriptive Statistics

Learning Objectives

In this chapter, you will learn how to calculate and report indicators of central tendency (mean, median, mode), variability (range, standard deviation, variance), and the shape of a distribution (skewness, kurtosis). You will also learn how to transform a set of raw scores to z scores.

CENTRAL TENDENCY AND VARIABILITY

Most of the analyses you will need to conduct can be found in the upper toolbar under the drop-down menu labeled 'Analyze.' Options to compute descriptive statistics and to analyze data using correlation, regression, *t* tests, and other commonly used statistics can be found there. We'll start with computing indicators of central tendency and variability.

Computing Indicators of Central Tendency and Variability
(Analyze→Descriptive Statistics→Frequencies)

Let's compute descriptive statistics on the weight variable we created in Chapter 1. Open the practice data file you created in Chapter 1 by clicking on the file name in your computer menu. Once the file is open, use the upper toolbar to go to **Analyze→Descriptive Statistics→Frequencies**.

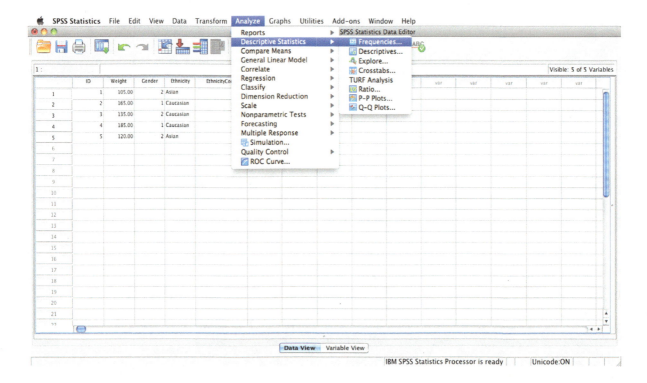

A 'Frequencies' dialog window like the one shown on the following page will now open. Highlight the variable **Weight in Pounds** shown on the left side of the window by clicking on it, and then press the **blue arrow** to move it into the box labeled **Variable(s)**.

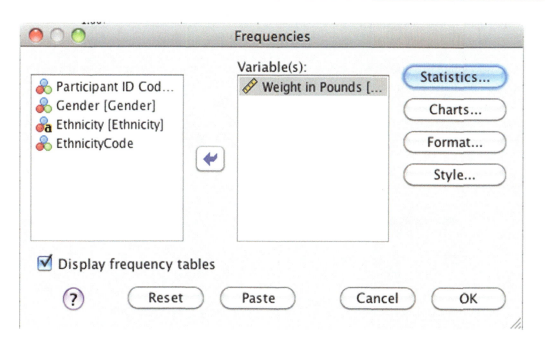

Next, press the **Statistics button** (highlighted in the image shown above). A 'Frequencies: Statistics' dialog window like the one shown below will now open. Check the boxes next to **Std. deviation**, **Variance**, **Range**, **Minimum**, **Maximum**, **Mean**, **Median**, and **Mode**.

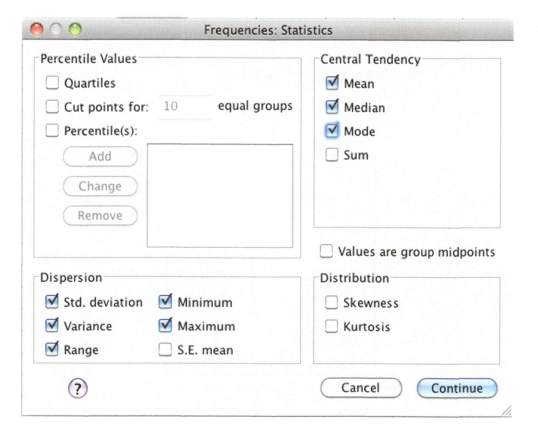

Press **Continue** and then **OK** to close the dialog windows.

Interpreting the Results

An output window like the one shown below will now appear. Whenever you analyze data in SPSS, the results will appear in an output window like this one.

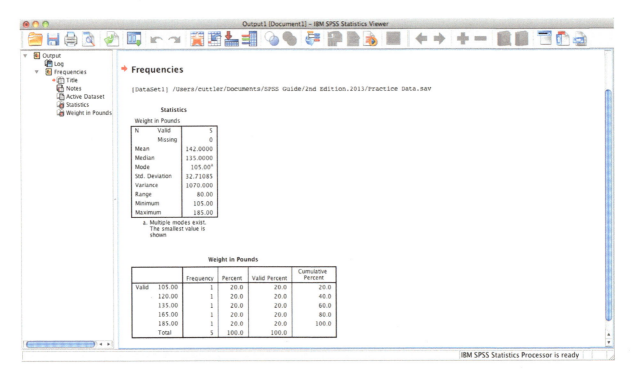

Let's take a closer look at the 'Statistics' table reproduced below.

Statistics

Weight in Pounds

N	Valid	5
	Missing	0
Mean		142.0000
Median		135.0000
Mode		105.00[a]
Std. Deviation		32.71085
Variance		1070.000
Range		80.00
Minimum		105.00
Maximum		185.00

a. Multiple modes exist. The smallest value is shown

As you can see, the table displays descriptive statistics for the variable 'Weight in Pounds.' Since we gave the variable this label when we defined its properties in Chapter 1, this label appears at the top of the table rather than the variable name. If we didn't label the variable, then the variable name 'Weight' would appear at the top of the table.

Sample Size (N)

The row labeled 'N' shows that the size of the sample is 5. The value of 5 next to the label 'Valid' indicates that we have data on the weights of 5 participants. The value of 0 next to the label 'Missing' indicates that we are not missing any participants' weight data. If you deleted the weight of one of the participants and re-ran the analysis, a value of 4 would appear next to 'Valid' and a value of 1 would appear next to 'Missing' to indicate that you are missing data on one participant's weight.

Variability

The remainder of the table displays the indicators of central tendency and variability that we requested. We will start with the indicators of variability, since they are the most straightforward to interpret. The row labeled 'Std. Deviation' shows the standard deviation is 32.71. In other words, the average distance of scores from the mean is 32.71 units (or lbs). The row labeled 'Variance' shows the variance is 1070.00 (the variance is just the standard deviation squared). The row labeled 'Range' shows the range of weights is 80.00; that is, the highest weight is 80 lbs higher than the lowest weight. The row labeled 'Minimum' shows that the lowest weight is 105.00 lbs, and the row labeled 'Maximum' shows that the highest weight is 185.00 lbs.

Central Tendency

Next, we will proceed to the indicators of central tendency. The row labeled 'Mean' shows the mean (i.e., average) weight is 142.00 lbs. The row labeled 'Median' reveals that the median (i.e., middlemost) weight is 135.00 lbs. It is important to note that the 'Statistics' table does not provide reliable information on modes. If the data contain one mode, the table will display its proper value. However, if there are multiple modes, the table will report only the lowest mode and it will display a note under the table (like the one shown on the preceding page) that indicates that multiple modes exist and only the smallest value is shown.

The mode is just the most frequently occurring score, so to find the other modes you can simply refer to the frequency table that also appears in the output window. This table is presented below.

Weight in Pounds

		Frequency	Percent	Valid Percent	Cumulative Percent
Valid	105.00	1	20.0	20.0	20.0
	120.00	1	20.0	20.0	40.0
	135.00	1	20.0	20.0	60.0
	165.00	1	20.0	20.0	80.0
	185.00	1	20.0	20.0	100.0
	Total	5	100.0	100.0	

The table clearly shows each value of the variable (each weight) in the unlabeled column. The column labeled 'Frequency' shows the frequency each of the values occurred (each of the weights occurred only one time, so 1 is listed in each cell). The column labeled 'Percent' shows the relative frequency using the entire sample (i.e., the percentage of participants with each score out of the total number of participants), the column labeled 'Valid Percent' shows the relative frequency using the participants who are not missing data (i.e., the percentage of participants with each score out of the number of participants with data on the variable)[4], and the column labeled 'Cumulative Percent' lists the cumulative percentages (the percentage of participants with each score or a score lower).

When two modes exist we say the distribution is bimodal and we report both modes, but when all of the scores in a distribution have the same frequency of occurrence we report that there is no mode. As shown above, the scores in the weight distribution all have the same frequency (a frequency of 1), so while the 'Statistics' table in the output window indicates that there are multiple modes, you would need to report that there is indeed no mode.

Reporting the Results

According to the APA style guide, most statistical notation should be italicized (the exception is Greek characters like μ and σ). An uppercase N is used to denote the size of the population or the total number of participants in a sample, while a lowercase n is typically used to denote the number of participants in a subsample (e.g., number of males in the sample).

According to the APA style guide, the mean of a population should be symbolized using μ, while the mean of a sample can be symbolized using either $\bar{X}$ or M. I will use M to denote the mean, but you should use the abbreviation your instructor uses and recommends. The statistical notation endorsed by the APA for the median is Mdn. The APA style guide does not provide an abbreviation for the mode.

The symbol σ is used to represent the standard deviation of a population and the symbol $σ^2$ represents the population variance. Either s or SD can be used to denote the standard deviation of a sample (once again you should choose the one recommended by your instructor) but only the notation s^2 should be used for the variance of a sample. The APA style guide does not provide an abbreviation for the range. Moreover, while technically the range is the highest score minus the lowest score, when we report the range using APA style, we simply report both the lowest and highest scores separated by a dash.

So, using APA style we would report the size of the sample as $N = 5$. If we wanted to report the number of males (a subsample), we could report $n = 2$. For the indicators of central tendency, we would report $M = 142.00$, $Mdn = 135.00$, and no mode. For the indicators of variability, we would report: $SD = 32.71$, $s^2 = 1070.00$, and range = 105.00 – 185.00. Note that since the words mode and range are not considered notation, they do not need to be italicized.

SKEWNESS AND KURTOSIS

Many inferential statistics tests carry the assumption that the data are normally distributed (i.e., symmetrical and bell shaped). Skewness and kurtosis statistics are used to characterize the shape of distributions and can therefore help you to determine whether your scores are indeed normally distributed.

The skewness statistic provides an indicator of the degree of asymmetry in your distribution. A skewness statistic of 0 would indicate a perfectly symmetrical distribution. Values above 0 (positive values) indicate

---

[4] Since we are not missing any weight data the Percent and Valid Percent columns display identical values.

that the distribution is positively skewed. As shown in the figure below, a positively skewed distribution contains more low scores (i.e., a peak on the left side of the distribution) and a longer right tail. Values below 0 (negative values) indicate that the distribution is negatively skewed. As shown in the figure below, a negatively skewed distribution contains more high scores (i.e., a peak on the right side of the distribution) and a longer left tail.

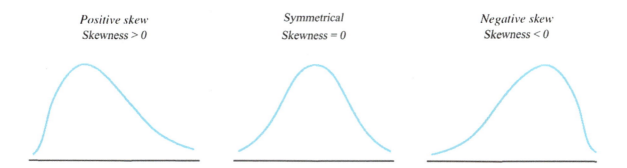

The kurtosis statistic provides an indicator of how tall and narrow the central peak is, and how fat (or elevated) the tails of the distribution are, relative to the normal distribution. A perfectly normal distribution has a kurtosis value of 3. However, many statistics programs, including SPSS, provide "excess kurtosis" values, which are equal to the standard kurtosis value minus 3. Thus, SPSS would report an excess kurtosis value of 0 for a perfectly normal distribution. Distributions with excess kurtosis values of 0 are referred to as *mesokurtic*. Distributions with excess kurtosis values above 0 (positive values) are referred to as *leptokurtic*. As shown in the image displayed below, leptokurtic distributions have a taller, narrower central peak and tails that are fatter (elevated). Distributions with excess kurtosis values below 0 (negative values) are referred to as *platykurtic*. As shown below, platykurtic distributions have a flatter central peak and tails that are shorter and thinner (remember: platy = flatty, not fatty).

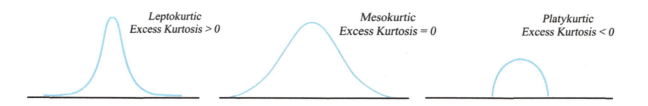

Computing Skewness and Kurtosis Statistics
(Analyze→Descriptive Statistics→Frequencies)

Skewness and kurtosis statistics are considered descriptive statistics, so they can also be computed by clicking **Analyze→Descriptive Statistics→Frequencies** in the upper toolbar. This will once again open the 'Frequencies' dialog window. The Weight variable should already appear in the Variable(s) box from the previous analyses (if it does not, then use the blue arrow to move it into this box). Next, click on the **Statistics button**. The 'Frequencies: Statistics' dialog window shown on the following page will now appear. Check the boxes next to **Skewness** and **Kurtosis** (you may wish to uncheck the remaining boxes to keep your output window less cluttered). Press **Continue** and then **OK** to close the dialog windows.

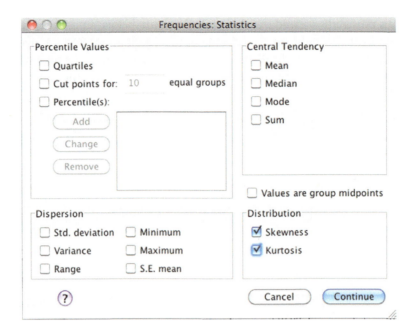

Interpreting the Results

The 'Statistics' table shown below will now appear in your output window.

Statistics

Weight in Pounds

N	Valid	5
	Missing	0
Skewness		.357
Std. Error of Skewness		.913
Kurtosis		-1.657
Std. Error of Kurtosis		2.000

The row labeled 'Skewness' provides the value of the skewness statistic. As you can see, it is 0.36. Similarly, the row labeled 'Kurtosis' displays an excess kurtosis statistic of –1.66. The table also presents the values of the standard errors for the skewness and excess kurtosis statistics in the rows labeled 'Std. Error of Skewness' and 'Std. Error of Kurtosis,' respectively.

As described earlier, perfectly normal distributions have skewness and excess kurtosis statistics of exactly 0. Since few things in life are ever perfect, these values will typically deviate from 0. The further the skewness and excess kurtosis values deviate from 0, the more the distribution deviates from normality. As a crude but general rule of thumb, if the skewness and excess kurtosis values are both between –1 and +1, the distribution can be considered normally distributed.[5] The skewness statistic of 0.36 shown in the table displayed above indicates that our distribution of weight scores is quite symmetrical. However, the excess kurtosis statistic of –1.66 indicates that the distribution is platykurtic and therefore deviates from normality.

---

[5] Some people use the less conservative range of –2 to +2 to indicate a normal distribution. Also note that there is a more sophisticated method for determining whether skewness and kurtosis statistics significantly deviate from 0 (this method involves dividing each statistic by its standard error), however this method is often criticized for being biased by sample size.

Reporting the Results

Skewness and kurtosis statistics are typically reported along with the value of their standard error in parentheses. The standard error is abbreviated *SE* (note the use of italics), and its values can exceed 1, so leading 0s before the decimal point are required. There are no abbreviations provided by the APA for skewness or kurtosis statistics and since the words skewness and kurtosis are not statistical notation, we will not italicize them. Taken together, we can simply report that for our weight variable, skewness = 0.36 (*SE* = 0.91) and excess kurtosis = −1.66 (*SE* = 2.00).

Z SCORES

Transforming Raw Scores to *z* Scores
(Analyze→Descriptive Statistics→Descriptives)

To transform a set of raw scores to *z* scores (i.e., standardized scores), use the upper toolbar to go to **Analyze→Descriptive Statistics→Descriptives**.

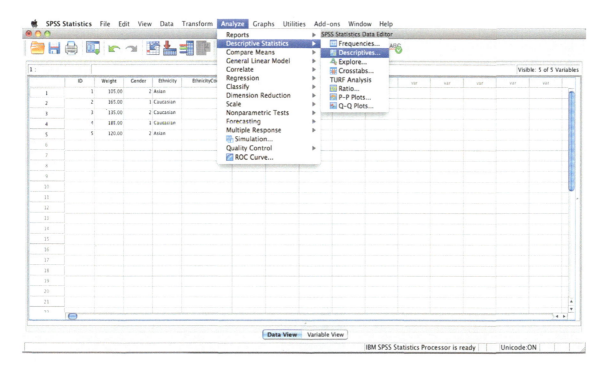

A 'Descriptives' dialog window like the one shown on the following page will now appear. Let's practice by transforming the weights into *z* scores. Move the weight variable into the **Variable(s) box** by highlighting **Weight in Pounds** and pressing the **blue arrow**. To obtain the *z* scores for this variable, simply check the box next to **Save standardized values as variables** (highlighted in the image shown on the following page). Press **OK**.

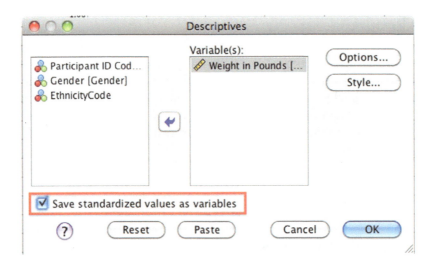

Interpreting the Results

An output window will appear showing descriptive statistics for the weight variable.[6] Since we are not currently interested in these statistics, you can simply return to the Data View window by clicking on the **red star icon** that appears in the icon toolbar of the output window (or by closing the output window).

As highlighted in the image below, a new column of scores labeled 'ZWeight' has been created. This column contains each participant's *z* score on the weight variable. For example, you can see that the participant with the ID code 1 received a *z* score of –1.13, indicating that she has a weight that is 1.13 standard deviations below the mean weight of 142.00 lbs. The participant with an ID code of 2 received a *z* score of 0.70, indicating that he has a weight that is 0.70 standard deviations above the mean.

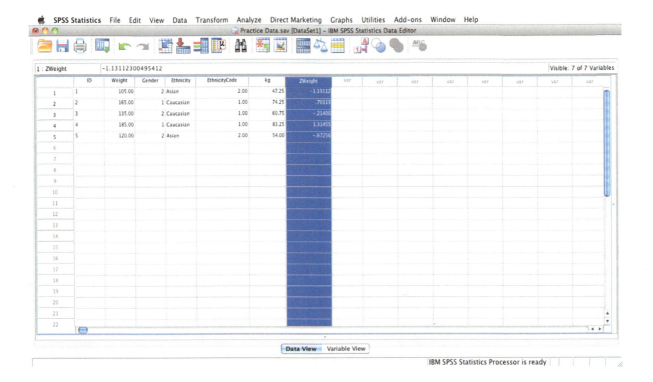

---

[6] This function can also be used to compute descriptive statistics however it does not provide the option to compute the median or mode.

Try On Your Own

Try combining some of the handy tools and tricks described in Chapter 1 with the analyses described in this chapter. For instance, try to find the mean weight of the males only (hint: select cases before computing the mean weight). Also try to calculate separate z scores for males and females (hint: split the file before creating the z scores), and then use the sort cases function to find the highest z score for each gender.

Correlation

Learning Objectives

In this chapter, you will learn how perform, interpret, and report the results of correlation analyses with variables measured on ratio, interval, ordinal, and nominal scales. You will also learn how to conduct, interpret, and report the results of one-tailed correlation and partial correlation analyses.

SAMPLE DATA FILES

The SPSS program includes a large number of sample data files. To follow the demonstrations in this chapter you will need to open a sample data file entitled 'Employee data'.

Opening Sample Data Files
(File→Open→Data)

Start by opening a blank SPSS worksheet. Once again, you can open the SPSS program by clicking on the program icon or filename (SPSSStatistics). When prompted, select **New Dataset** and then click **OK**. Next, go to **File→Open→Data**. Alternatively, you can simply click on the orange folder icon in the icon toolbar.

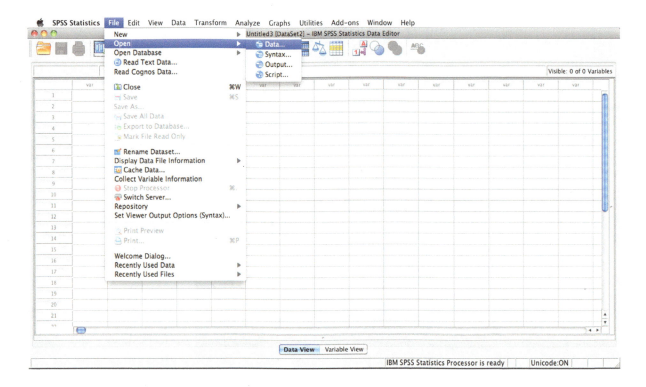

A dialog window labeled 'Open Data' will now appear. You will need to use this dialog window to find the location of your SPSS program files. If you used the default installation, the SPSS program and the sample data files will be located in your applications (Mac) or program (PC) files in a folder labeled 'IBM.' Find and open your **Applications** or **Program Files**, then find and open **IBM→SPSS→Statistics→22→Samples**. As shown on the following page, a series of sample SPSS data files should now appear. Find and highlight the file entitled '**Employee data**' and click **Open**.

If no files appear when you click on Samples then close SPSS and using your computer menu go to your Applications or Program Files and find **IBM→SPSS→Statistics→22→Samples**. The series of sample data files should then appear. Double click on the file name '**Employee data**' to open the file.

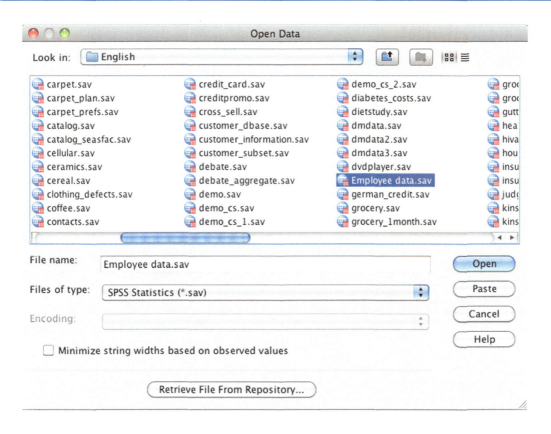

The employee data file contains data on 474 participants ($N = 474$). Whenever you open a data file that you did not create you should immediately examine the Variable View window to familiarize yourself with the variables, their scales of measure, and their values. Take a minute now to go to the **Variable View** window so that you can familiarize yourself with this data file. By looking at the variable names and labels, you will see that the file contains data on each participant's gender, birth date (named 'bdate'), years of education (named 'educ'), type of job (named 'jobcat'), current salary in dollars (named 'salary'), beginning salary in dollars (named 'salbegin'), number of months in the job (named 'jobtime'), number of months of previous experience (named 'prevexp'), as well as information on whether the individual is a minority (named 'minority'). The Values column shows that individuals working in a clerical position were coded with a 1, those working in a custodial position were coded with a 2, and those working in a management position were coded with a 3. Finally, this column shows that minorities were coded with a 1, and non-minorities were coded with a 0.

SCATTERPLOTS

It is only appropriate to calculate a correlation coefficient when the variables you plan to correlate show a linear relationship. This is because the analysis involves the assumption that the relationship between the variables is linear and the magnitude of the correlation coefficient will be underestimated if the variables show a curvilinear relationship. Thus, before conducting any correlation analysis you should first create a scatterplot to ensure that the variables show a linear relationship.

Generating Scatterplots
(Graphs→Legacy Dialogs→Scatter/Dot)

We will begin by creating a scatterplot to examine the relationship between current salary and beginning salary. To create the scatterplot, you will need to use the upper toolbar to go to **Graphs→Legacy Dialogs→Scatter/Dot.**

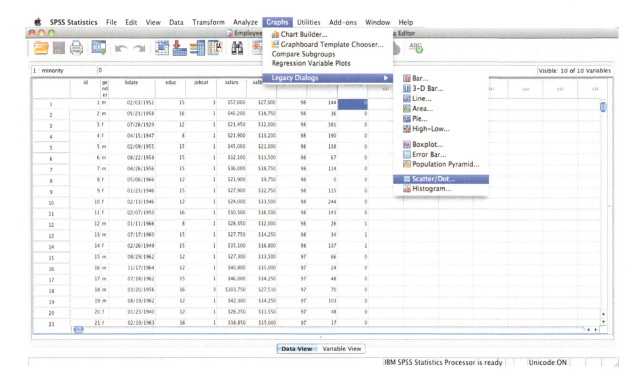

A 'Scatter/Dot' dialog window like the one displayed below will appear. Click on the **Simple Scatter** window and then click **Define**.

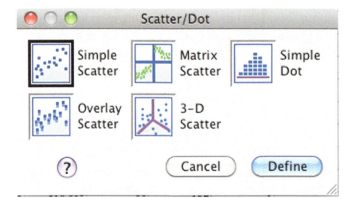

Next, a 'Simple Scatterplot' dialog window like the one shown on the following page will appear with all of the variables in the dataset listed on the left side. Highlight the variable **Current Salary** by clicking on it, and move it over to the **Y Axis box** by clicking on the corresponding **blue arrow**. Next, highlight the variable **Beginning Salary** by clicking on it, and move it over to the **X Axis box** by clicking on the corresponding **blue arrow**. Click **OK** to close the dialog window.

An output window will now appear displaying the scatterplot presented below. The scatterplot clearly shows that the relationship between current salary and beginning salary is linear. Since the relationship is linear, it is appropriate to calculate the correlation coefficient.

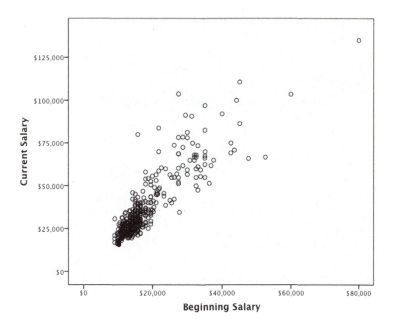

Scatterplots will not usually look quite as perfect as the one shown above. When one of the variables is nominal with two levels/values (e.g., gender), scatterplots can be particularly strange looking and difficult to interpret (the data points will all fall on two separate lines because there are only two levels/values of the nominal variable). In most cases, you should assume the relationship is linear unless the scatterplot depicts a clear curvilinear relationship.

Try On Your Own

Practice creating some scatterplots on your own. Try to create scatterplots depicting the relationships between educational level and current salary, as well as between educational level and beginning salary. You will see those relationships are not as clear-cut as the relationship we examined between current salary and beginning salary. Since educational level is defined as an ordinal variable and contains a limited number of levels, the data points all fall on separate lines. However, since the scatterplots show that there are no clear curvilinear relationships, we will proceed to compute the correlation coefficients later in this Chapter.

PEARSON CORRELATION COEFFICIENTS (r)

Correlation coefficients are most commonly computed using two variables measured on interval or ratio scales. This type of correlation coefficient is formally referred to as a Pearson correlation coefficient (or even more formally a Pearson product-moment correlation coefficient) and is symbolized as r.

Computing Pearson Correlation Coefficients
(Analyze→Correlate→Bivariate)

Let's start by considering the correlation between current salary and beginning salary, each of which is measured on a ratio scale. To compute the Pearson correlation coefficient for these variables, you will need to use the upper toolbar to go to **Analyze→Correlate→Bivariate**.

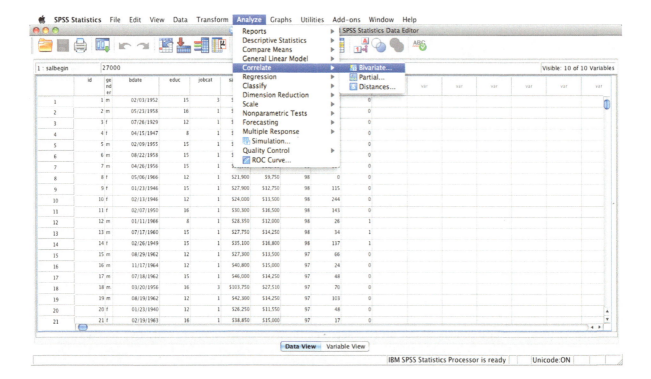

As shown below, a dialog window labeled 'Bivariate Correlations' will now open with all of the variables in the dataset listed on the left side.[7] Click on **Current Salary** and use the **blue arrow** to move it into the **Variables box**. Next, click on **Beginning Salary** and use the **blue arrow** to move it into the **Variables box**. Press **OK** to close the dialog window and initiate the analysis.

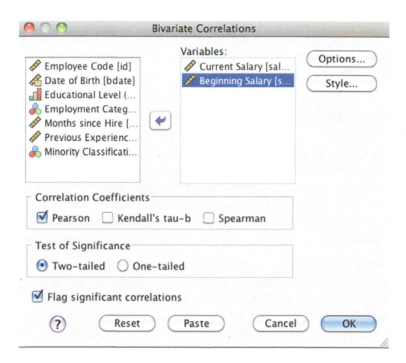

Interpreting the Results

The following 'Correlations' table will now appear in the output window.

Correlations

		Current Salary	Beginning Salary
Current Salary	Pearson Correlation	1	.880[**]
	Sig. (2-tailed)		.000
	N	474	474
Beginning Salary	Pearson Correlation	.880[**]	1
	Sig. (2-tailed)	.000	
	N	474	474

**. Correlation is significant at the 0.01 level (2-tailed).

The rows labeled 'N' indicate how many pairs of scores (participants) were included in the analysis. If you were missing the beginning salary of one participant, a value of 473 would appear in these rows instead of 474 because SPSS will only include participants who have data on both of the variables.

---

[7] Since the gender variable is a string variable (i.e., a variable that contains text), it is not contained in the list. Remember, SPSS will not analyze variables entered using text.

The rows labeled 'Pearson Correlation' contain the Pearson correlation coefficients. You can see that the correlation between current salary and current salary is 1, and the correlation between beginning salary and beginning salary is also 1. The correlation between a variable and itself will always equal 1, so this is not terribly interesting or informative. Of primary interest, the table shows that the correlation between current salary and beginning salary is .88.

Correlation coefficients of .10 are generally interpreted as small, correlations of .30 are generally interpreted as moderate, and correlations of .50 and above are generally interpreted as large. Thus, our correlation of .88 indicates that there is a large positive correlation between beginning salary and current salary; that as beginning salary increases, so does current salary.

Determining Statistical Significance

In addition to knowing the size and direction of the correlation between two variables, it is also useful to know whether the correlation coefficient is statistically significant. Statistically significant effects are those that have a low probability of occurring due to chance alone. Specifically, an effect is considered statistically significant if there is a low probability that it would be found if there was no real effect in the population. In the context of correlation, a correlation coefficient is considered statistically significant if there is a low probability that it would be found if the actual correlation in the population was 0.

Typically, to be considered statistically significant, the probability that an effect is due to chance alone must be .05 or less. Although sometimes researchers set a more stringent criterion and require that this probability be .01 or less before they will consider the effect statistically significant. This criterion or threshold is referred to as the alpha level and it must be decided upon prior to conducting the study. Once again, alpha is a threshold we set prior to conducting a study representing our willingness to conclude that there is an effect (e.g., that there is a correlation) when in reality there is no effect in the population and our results are just due to chance. And because we don't want to make this kind of error very frequently, this threshold is always set low (typically either at .05 or .01).

Information about whether a correlation coefficient is statistically significant can be found in the rows of the 'Correlations' table labeled 'Sig. (2-tailed)'. The p values, or significance levels, provided in these rows reflect the probability that we would obtain our result, or a result more extreme, if there is no real effect in the population. In other words, it is the probability that chance alone is operating and in reality the correlation in the population is 0. If the p value listed in the 'Sig. (2-tailed)' row is less than or equal to the alpha level set prior to conducting the study, the correlation is considered statistically significant. If the p value in the 'Sig. (2-tailed)' row is greater than the alpha level set, the correlation is not statistically significant. To use the analogy of the limbo dance, you can think of the alpha level as a limbo bar (the threshold we set), and the p value as the person (the value we want to get under alpha). When p is under alpha, our effect is statistically significant (and it's time to party!).

As described above, the alpha level is typically set at .05, and as such, that is the alpha level we will use (again this decision really should have been made before we ran the analysis, but at that point I hadn't introduced the concept to you). The 'Correlations' table displays the p value for the correlation between current salary and beginning salary as .000. This value

indicates that there is less than a .001 chance that this correlation, or one larger, would be found if only chance is operating and in reality there is no correlation between these variables. Since this *p* value is less than .05 (the alpha level we set), we can conclude that there is a statistically significant correlation between the variables. The stars next to the correlation coefficients in the table also indicate that the correlation is significant. If stars were not present, it would indicate that the correlation is not statistically significant.

Reporting the Results

The results of correlation analyses are reported using the following format: $r(df) = .\#\#, p = .\#\#$. Note that the *r* and *p* values do *not* have leading 0s before the decimal point because these values can never exceed a value of 1 (see the section Reporting Decimal Remainders and Rounding in Chapter 1 for a complete description of reporting decimal remainders). Also recall that APA style requires us to italicize all statistical notation (except Greek characters), so *r* and *p* should both be italicized.

We have already established that *r* is used to symbolize the Pearson correlation coefficient and *p* symbolizes the *p* value or significance level. Now all you need to know is that *df* stands for degrees of freedom. The number of degrees of freedom is not provided in the output table, but the degrees of freedom for correlations are very easy to calculate: they simply equal the size of the sample minus 2 ($df = N - 2$). As reviewed above, the sample size is 474, so our degrees of freedom are: 474 – 2 = 472. Degrees of freedom are discrete numbers, so no decimal remainder should ever be reported.

Therefore, for the results obtained above we could report the following:

> There is a large positive correlation between current salary and beginning salary that is statistically significant, $r(472) = .88, p < .001$.

Try On Your Own

Try to compute the Pearson correlation between current salary and months since hire. Use the conventional alpha level of .05 and practice reporting the results using APA style.

SPEARMAN RANK ORDER CORRELATION COEFFICIENTS (r_S)

Correlation coefficients can also be computed with variables measured on ordinal scales. When one of the variables is measured on an ordinal scale and the other variable is measured on an ordinal, interval, or ratio scale the correlation is referred to as a Spearman rank order correlation coefficient, or Spearman's rho (pronounced row), and it is symbolized as r_s.

Computing Spearman Rank Order Correlation Coefficients
(Analyze→Correlate→Bivariate)

Let's begin by computing some Spearman rank order correlations using the variable educational level, which has been defined as an ordinal variable in this dataset[8] (see Variable View). We will correlate educational

---

[8] Frankly, I disagree with this classification of years of education as an ordinal variable (I think it should be coded as a Scale variable), but I will go along with it for the pedagogical purpose of illustrating a Spearman rank order correlation.

level with both beginning salary and current salary. We will stick with the convention and set alpha at .05.[9] Go to **Analyze→Correlate→Bivariate**. Put **Current Salary**, **Beginning Salary**, and **Educational Level** in the **Variables box** by highlighting each and clicking on the **blue arrow**. Do **NOT** check the box labeled 'Spearman' (see the Important Note box below for an explanation). Click **OK** to initiate the analysis.

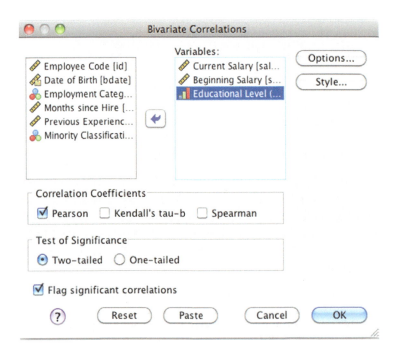

Important Note

While the option to calculate a Spearman correlation coefficient is provided in the 'Bivariate Correlations' dialog window shown above, you should **avoid using this option**. The formula for calculating Spearman's rho is merely a simplified version of Pearson's formula. If there are no ties on the ordinal variables, or if there are ties that have been handled properly, then the value of Spearman's rho will be identical to the value of the Pearson correlation coefficient. When the option to calculate Spearman's rho is checked, SPSS will automatically rank order the data for *both* of the variables (rather than just the variable defined as ordinal). For this reason, you should never use the option to compute a Spearman correlation coefficient unless both of the variables you want to correlate were measured on ordinal scales. Even in this case, the value of the Pearson correlation will be identical to the value of Spearman's rho (if ties are properly handled[1]), so many researchers will still calculate the correlation between two ordinal variables using the Pearson option.

[1] Ties on ordinal variables should be handled in a specific way, by assigning the mean rank of the tied variables (e.g., if there is a tie for the rank of 2, both rankings should be changed to 2.5 and the ranking of 3 should be skipped). Ties in the educational level variable have not been properly handled in this dataset. The option to properly rank order the data is available in SPSS (Transform→Rank Cases). Since the mishandling of ties has a trivial effect on the magnitude of the correlation coefficients (properly ranking the data increases each correlation by less than .02), we will not worry about ranking the data.

---

[9] Remember this means if the results reveal a *p* value of .05 or lower we will consider the correlation statistically significant.

Interpreting the Results

The following table will now appear in the output window. It shows all of the correlations between the three variables (in the rows labeled 'Pearson Correlation'), as well as the associated p values (in the rows labeled 'Sig. (2-tailed)'), and sample sizes (in the rows labeled 'N'). Specifically, the bottom portion of the table shows that there are large positive correlations between educational level and current salary (.66) and educational level and beginning salary (.63) that are statistically significant ($p < .001$). Remember, we will need to compute the degrees of freedom ourselves by subtracting 2 from N. Since $N = 474$, our degrees of freedom are: $474 - 2 = 472$.

Correlations

		Current Salary	Beginning Salary	Educational Level (years)
Current Salary	Pearson Correlation	1	.880**	.661**
	Sig. (2-tailed)		.000	.000
	N	474	474	474
Beginning Salary	Pearson Correlation	.880**	1	.633**
	Sig. (2-tailed)	.000		.000
	N	474	474	474
Educational Level (years)	Pearson Correlation	.661**	.633**	1
	Sig. (2-tailed)	.000	.000	
	N	474	474	474

**. Correlation is significant at the 0.01 level (2-tailed).

Reporting the Results

Based on these results, we can report the following:

> There are large positive correlations between current salary and educational level, $r_s(472) = .66$, $p < .001$, and between beginning salary and educational level, $r_s(472) = .63$, $p < .001$, that are statistically significant.

POINT BISERIAL (r_{pb}) AND PHI COEFFICIENTS (ϕ)

Correlations can also be computed with variables measured on nominal scales as long as the nominal variables are dichotomous (they contain only two categories). Correlations should never be computed with variables measured on nominal scales that contain more than two categories because the results would be meaningless (e.g., the correlation between undergraduate major and hair color could not be interpreted in any meaningful way).

When one variable is measured on a nominal scale and the other variable is measured on an interval or ratio scale the correlation is referred to as a point biserial correlation and it is symbolized as r_{pb}. When both of the variables are measured on nominal scales the correlation is referred to as a phi coefficient and it is symbolized using the Greek character ϕ.

Computing Point Biserial Correlation Coefficients
(Analyze→Correlate→Bivariate)

We will practice computing point biserial correlations by correlating gender with both current salary and beginning salary. We will stick with the convention and set alpha at .05.

Since gender is entered as a string (text) variable, we will not be able to perform any analyses with it until it has been recoded using a numeric code. Go to **Transform→Recode into Different Variables**. Label the recoded variable **GenderCode** and then proceed to recode the current gender code **m** to **1** and **f** to **2** (refer to the section Recoding Variables on pages 16–18 if you forget how to do this). Save this data file since we will use this file with the numerically recoded gender variable in subsequent chapters.

Now that we have a numerically coded gender variable we can compute correlations with it. Point biserial correlations and phi coefficients are computed in the same way as Pearson correlation coefficients,[10] so using the upper toolbar simply go to **Analyze→Correlate→Bivariate**. Put **Current Salary**, **Beginning Salary**, and **GenderCode** in the **Variables box** by highlighting each and clicking on the **blue arrow**. Click **OK** to initiate the analysis.

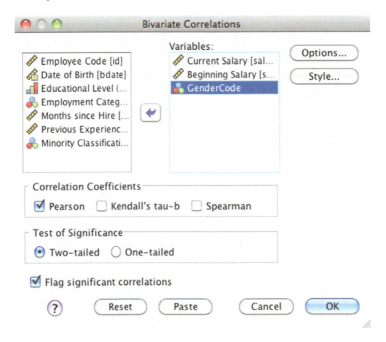

Interpreting the Results

The following 'Correlations' table will now appear in the output window.

Correlations

		Current Salary	Beginning Salary	GenderCode
Current Salary	Pearson Correlation	1	.880**	-.450**
	Sig. (2-tailed)		.000	.000
	N	474	474	474
Beginning Salary	Pearson Correlation	.880**	1	-.457**
	Sig. (2-tailed)	.000		.000
	N	474	474	474
GenderCode	Pearson Correlation	-.450**	-.457**	1
	Sig. (2-tailed)	.000	.000	
	N	474	474	474

**. Correlation is significant at the 0.01 level (2-tailed).

---

[10] While there are different formulas for calculating point biserial and phi coefficients, they are merely simplified versions of Pearson's formula and they produce the same results as Pearson's formula. Since SPSS is doing all of our work, we don't need to worry about simplified formulas!

The bottom portion of the table shows that there are moderately sized negative correlations between gender and current salary (−.45) and gender and beginning salary (−.46) that are statistically significant ($p < .001$). Before we can fully interpret these correlations we need to consider the code used for the gender variable. We assigned males a code of 1 (a lower value) and females a code of 2 (a higher value). The negative coefficients indicate that high values on one variable are associated with low values on the other variables. As such, we can conclude that being female (a higher value) is associated with significantly lower beginning salaries, as well as with significantly lower current salaries, or similarly that being male (a lower value) is associated with significantly higher beginning and current salaries.

Reporting the Results

Based on these results, we can report the following:

> There are moderately sized negative correlations between gender and current salary, $r_{pb}(472)$ = −.45, $p < .001$, and between gender and beginning salary, $r_{pb}(472) = −.46$, $p < .001$, that are statistically significant. These correlations indicate that being female is associated with earning significantly lower beginning and current salaries.

Computing Phi Coefficients
(Analyze→Correlate→Bivariate)

Finally, we will compute the phi coefficient for the relationship between minority classification and gender. We will stick with the convention and set alpha at .05. Note that we can only use minority classification as a variable in our analysis because it has been coded in a binary manner and therefore only contains two categories (minority and nonminority).[11] Remember, meaningful correlations cannot be computed with nominal variables that contain more than two categories. To compute the phi coefficient you will once again need to go to **Analyze→Correlate→Bivariate**. Put **Minority Classification** and **GenderCode** in the **Variables box** by highlighting each and clicking on the **blue arrow**. Click **OK** to initiate the analysis.

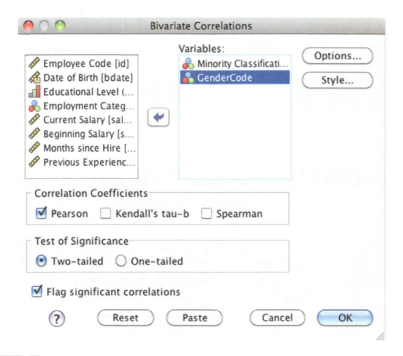

---

[11] You may also notice that a code of 9 was used for missing data. Since there are no missing data on this variable (go ahead and check for yourself using Analyze→Descriptive Statistics→Frequency), it is a true dichotomous variable and can be analyzed using correlation.

Interpreting the Results

The following 'Correlations' table will now appear in the output window.

Correlations

		Minority Classification	GenderCode
Minority Classification	Pearson Correlation	1	-.076
	Sig. (2-tailed)		.100
	N	474	474
GenderCode	Pearson Correlation	-.076	1
	Sig. (2-tailed)	.100	
	N	474	474

Since the p value displayed in the table is higher than our alpha level of .05, we must conclude that the correlation is not statistically significant and therefore that there is no correlation between the variables.

Nevertheless, it is good to practice interpreting the direction of phi coefficients. Before you can interpret any phi coefficient you need to consider the codes used for both variables. As we just reviewed, we assigned males a code of 1 (a lower value) and females a code of 2 (a higher value). By looking in the Values column in the Variable View window you will find that individuals who are not minorities were labeled with a 0 (a lower value), while those who are minorities were labeled with a 1 (a higher value). The presence of a negative coefficient indicates that higher values on one variable tend to be associated with lower values on the other variable. Thus, it appears that being female is associated with not being a minority, and that being male is associated with being a minority. Once again, however, this relationship is not statistically significant (or terribly interesting!).

Reporting the Results

Note that phi is a Greek character, and as such, the symbol should not be italicized. Based on these results, we could report the following:

There is a small negative correlation between minority classification and gender, $\phi(472) = -.08$, $p = .10$, that is not statistically significant.

DIRECTIONAL HYPOTHESES AND ONE-TAILED CORRELATION ANALYSES

In the previous examples, we considered only non-directional hypotheses (we examined whether correlations existed, but we did not predict the direction of effects). However, we could have instead used directional hypotheses in which we predicted the direction of the effects. When our hypothesis is non-directional we say the statistical analysis is two-tailed (we consider two possible outcomes: a positive effect *and* a negative effect). In contrast, when our hypothesis is directional we say the statistical analysis is one-tailed (we consider only one possible outcome: a positive *or* a negative effect).

In the Try On Your Own box on page 41, you were asked to practice computing the Pearson correlation between current salary and months since hire. You should have found that the correlation between these variables is small, positive, and not statistically significant, $r(472) = .08$, $p = .07$. Since there is no reason to believe that higher current salaries would be associated with less experience (fewer months since hire), we could have instead made a directional hypothesis. Specifically, we could have made the directional hypothesis that there is a positive correlation between current salary and the number of months since hire. Let's assume we did that. Once again, we will stick with convention and set alpha at .05 (one-tailed).

Conducting One-Tailed Correlation Analyses
(Analyze→Correlate→Bivariate)

In order to compute a one-tailed correlation analysis you once again need to go to **Analyze→Correlate→Bivariate**. This will open the 'Bivariate Correlations' dialog window. Move the variables **Current Salary** and **Months since Hire** into the **Variables box** using the **blue arrow**. To change the default from a two-tailed analysis to a one-tailed analysis, simply click on the **One-tailed** option (highlighted in the image shown below). Finally, click **OK** to execute the analysis.

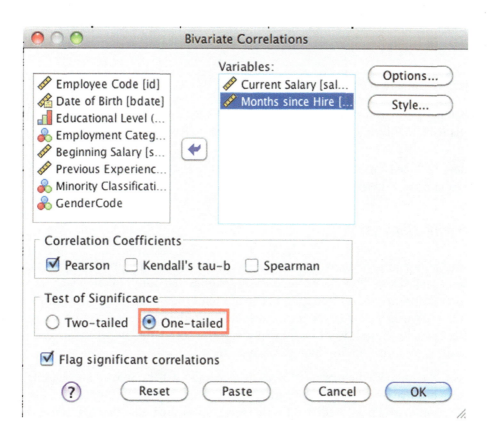

Interpreting the Results

The following table will now appear in the output window.

Correlations

		Current Salary	Months since Hire
Current Salary	Pearson Correlation	1	.084*
	Sig. (1-tailed)		.034
	N	474	474
Months since Hire	Pearson Correlation	.084*	1
	Sig. (1-tailed)	.034	
	N	474	474

*. Correlation is significant at the 0.05 level (1-tailed).

As you can see, the value of the correlation coefficient does not change, it is still $r(472) = .08$. The only value that changes with a one-tailed analysis is the p value. First, you should see that the rows containing the p values are now labeled 'Sig. (1-tailed)' to indicate that they are p values for a one-tailed test. The p value for a one-tailed test will always be exactly half the value it would be for a two-tailed test. Recall the p value for the two-tailed correlation between these variables was .067. The p value for this one-tailed correlation is now half that size, .034. Since .03 is less than our alpha of .05, this one-tailed correlation is statistically significant! But wait, before making any conclusions we need to check to make sure that the correlation is in the same direction as we predicted, and indeed it is in the positive direction.

Reporting the Results

Using APA style, we can now report the following:

> There is a small positive correlation between current salary and number of months since hire that is statistically significant, $r(472) = .08$, $p = .03$ (one-tailed).

Note that we have included 'one-tailed' in parentheses next to the p value to indicate that our hypothesis was directional and our p value is one-tailed.

Predicting the Wrong Direction

It is important to note that SPSS does not know which direction we are predicting when we select the option to run a one-tailed test. SPSS always generously assumes that our prediction is in the correct direction (the direction the results end up falling). As such, it is important to consider how things would change if we predicted the wrong direction (if the results turned out in the opposite direction we predicted). Let's assume our directional hypothesis was that there is a *negative* correlation between current salary and the number of months since hire (a foolish hypothesis indeed).

If we were to run the one-tailed correlation analysis we would get the same 'Correlations' table shown above indicating a positive correlation of .08 and a p value of .03. However, since the results turned out in the opposite direction to which we predicted (in this scenario we are predicting a negative correlation but the results show a positive correlation), we would need to adjust the p value by subtracting it from 1. For added precision we should use a one-tailed p value that has been rounded to four decimal places.[12]

$$p = 1 - .0337$$
$$p = .97$$

---

[12] Remember you can reveal the full decimal remainder by clicking repeatedly on the value in the output window.

Thus, our reported results would be: $r(472) = .08$, $p = .97$ (one-tailed). Since in this scenario $p > .05$, we would need to conclude that there is not a significant negative correlation between current salary and months since hire. Note: in this scenario we *cannot* conclude that there is a significant positive correlation between the variables (see below box). This is the risk we take when we make a directional hypothesis. If the results turn out to be significant in the opposite direction, we must adjust the p value and conclude the result is not significant. However, with this risk comes greater power to find an effect in the predicted direction (because our p values are cut in half).

Important Note

It is considered unethical to change a non-directional hypothesis to a directional hypothesis after discovering that the effect is significant for a one-tailed test but not a two-tailed test. It would be comparable to changing alpha from .05 to .10 after discovering that your p value is .09. It is even worse to change the direction of your directional hypothesis after discovering that the results are in the opposite direction to which you predicted!

PARTIAL CORRELATION ($r_{ab.c}$)

Overview

Correlation does not permit determination of causation. The third variable problem is one reason why we can never infer causation based on the results of correlation. A correlation between two variables may appear simply because both of the variables are related to some extraneous third variable. For instance, if we assessed 100 participants and discovered the correlation between their depression levels and the number of fast food meals they consumed last month was $r(98) = .35$, $p < .001$, we could not conclude that consuming more fast food causes people to become more depressed, or that higher levels of depression cause people to consume more fast food, because a third variable may be responsible for the relationship. In this case, reduced income may be related to more fast food consumption and higher levels of depression, rather than a direct causal relationship between fast food consumption and depression existing.

The only way to determine causality is through the use of the experimental method because potential confounding (i.e., third) variables are physically controlled. While we cannot physically control for third variables using correlation, we can statistically control for them. We can rule out third variables using a technique called partial correlation. This technique allows us to examine the relationship between two variables (e.g., fast food consumption and depression) after statistically controlling for a potential third variable (e.g., income).[13]

---

[13] Note that even if we use partial correlation to rule out third variables, we still cannot determine causation. This is because other third variables may still be at play, and because of the directionality problem.

The symbol provided in the APA style guide for the partial correlation coefficient is $r_{ab.c}$. The subscripts 'a' and 'b' refer to the two variables you are correlating and the subscript '.c' refers to the variable you are partialling out or controlling for. So, for this example the subscripts 'a' and 'b' refer to depression levels and fast food consumption, and the '.c' refers to income.

Interpreting Partial Correlations

Interpreting a partial correlation simply involves comparing the magnitude of the original correlation coefficient with the magnitude of the partial correlation coefficient. The original correlation coefficient is often referred to as the bivariate correlation coefficient ('bi' indicates two and 'variate' indicates variables, so the bivariate correlation is the relationship between two variables). When the bivariate correlation coefficient is large and statistically significant and the partial correlation coefficient is substantially lower and is not statistically significant, it suggests that the variable that was statistically partialled out is a third variable that is responsible for the bivariate correlation. To illustrate, assume you ran a correlation analysis and found that the correlation between depression and fast food consumption was, $r(98) = .35$, $p < .001$. Now assume you ran a partial correlation analysis—correlating depression with fast food consumption, after controlling for income—and found that the partial correlation was, $r_{ab.c}(97) = .09$, $p = .37$. The large difference between these coefficients and the drop from statistical significance suggests that income is responsible for the correlation between depression and fast food consumption. In other words, this outcome suggests that there is little to no relationship between depression and fast food consumption when income is statistically controlled.

When the bivariate correlation coefficient and the partial correlation coefficient are identical or very similar, it suggests that the variable that was partialled out is not a third variable, that it is having little influence on the bivariate correlation. To illustrate, if the bivariate correlation between depression and fast food consumption was, $r(98) = .35$, $p < .001$, and the partial correlation was, $r_{ab.c}(97) = .32$, $p = .002$, then we could conclude that income does not account for the correlation between depression and fast food consumption. This is because controlling for income had very little effect on the magnitude of the correlation.

When the partial correlation coefficient is statistically significant but substantially lower than the bivariate correlation coefficient, it suggests that the variable that was partialled out is a third variable, but that there is also a relationship between the variables of interest that is independent of the third variable. In other words, if a significant relationship still exists between the variables after the third variable is controlled, it suggests that a relationship exists between the variables of interest that is independent of the variable that was controlled. However, because controlling for the third variable reduced the magnitude of the correlation, we can infer that the third variable was increasing the magnitude of the bivariate correlation. For example, if the bivariate correlation between depression and fast food consumption was, $r(98) = .35$, $p < .001$, and the partial correlation was, $r_{ab.c}(97) = .20$, $p = .05$, you would know that income accounts for *some* of the relationship between depression and fast food consumption (because controlling for income decreased the correlation coefficient by .15), but that there is still a relationship between depression and fast food consumption that is independent from income (the partial correlation is still statistically significant).

Computing Partial Correlations
(Analyze→Correlate→Partial)

Let's try to calculate the partial correlation between current salary and educational level, controlling for beginning salary.[14] Go to **Analyze→Correlate→Partial**.

---

[14] Technically, we shouldn't be using an ordinal variable in a partial correlation analysis, but once again I disagree with the categorization of educational level (in years) as ordinal, so we're going to go ahead and do it anyway (feel free to call the stats police on me).

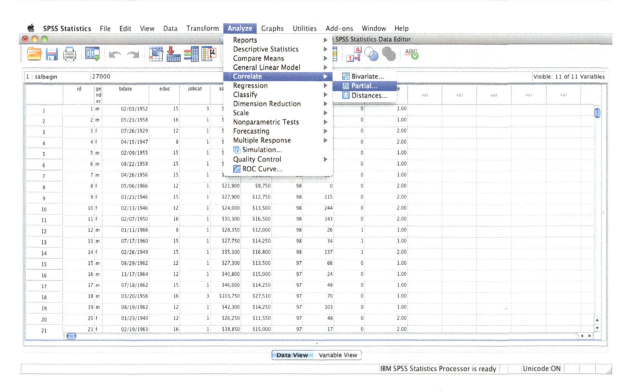

A 'Partial Correlations' dialog window will now appear. Move **Current Salary** and **Educational Level** into the **Variables box** using the corresponding **blue arrow**, and move **Beginning Salary** into the **Controlling for box** using the corresponding **blue arrow**.

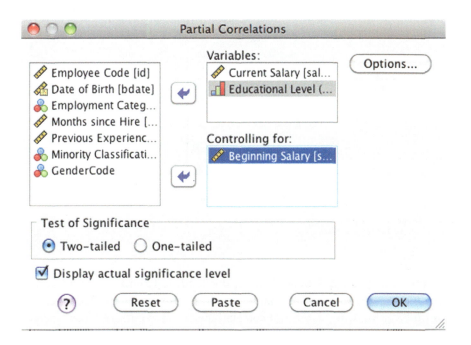

Next, click on the **Options button** in the 'Partial Correlations' dialog window. This will open the 'Partial Correlations: Options' dialog window shown on the following page. **Check** the **Zero-order correlations box**. By checking this option, the output window will contain both the partial and the bivariate correlations (which SPSS refers to as zero-order correlations), allowing us to more easily compare their values. Click **Continue** and then **OK** to close the dialog windows and initiate the analysis.

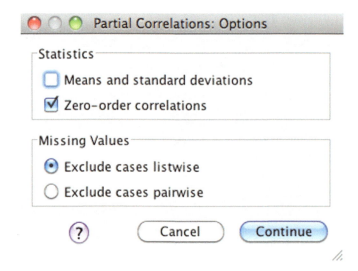

Interpreting the Results

The following table will appear in your output window.

Correlations

Control Variables			Current Salary	Educational Level (years)	Beginning Salary
−none−[a]	Current Salary	Correlation	1.000	.661	.880
		Significance (2−tailed)	.	.000	.000
		df	0	472	472
	Educational Level (years)	Correlation	.661	1.000	.633
		Significance (2−tailed)	.000	.	.000
		df	472	0	472
	Beginning Salary	Correlation	.880	.633	1.000
		Significance (2−tailed)	.000	.000	.
		df	472	472	0
Beginning Salary	Current Salary	Correlation	1.000	.281	
		Significance (2−tailed)	.	.000	
		df	0	471	
	Educational Level (years)	Correlation	.281	1.000	
		Significance (2−tailed)	.000	.	
		df	471	0	

a. Cells contain zero-order (Pearson) correlations.

The top portion of the table (that lists '−none−ᵃ' under the heading 'Control Variables') shows the bivariate correlations between current salary, educational level, and beginning salary. These results are consistent with our previous findings, showing significant correlations between current salary and educational level, $r_s(472) = .66, p < .001$, current salary and beginning salary, $r(472) = .88, p < .001$, and beginning salary and educational level, $r_s(472) = .63, p < .001$.

The bottom portion of the table (that lists 'Beginning Salary' under the heading 'Control Variables') shows the partial correlation between current salary and educational level, controlling for beginning salary. As highlighted in the table, the partial correlation is statistically significant, $r_{ab.c}(471) = .28, p < .001$.

For some reason, when the partial correlation option is used to analyze data, degrees of freedom are provided in the table (rather than the sample size) and stars are not placed next to significant coefficients. So, while you will not have to hand calculate the degrees of freedom,[15] you will have to rely on the p values (rather than stars) to determine statistical significance.

By comparing the bivariate correlation between current salary and educational level, $r_s(472) = .66$, $p < .001$, with the partial correlation between current salary and educational level, after controlling for beginning salary, $r_{ab.c}(471) = .28$, $p < .001$, we can infer that beginning salary accounts for some, but not all of the correlation between current salary and educational level. By controlling for beginning salary, the magnitude of the correlation dropped by .38 (.66 − .28 = .38), which is substantial. However, the partial correlation between current salary and educational level is still moderate in size and statistically significant, suggesting that there is a correlation between current salary and educational level that is independent of beginning salary.

Reporting the Results

The following results could be reported:

> The partial correlation between current salary and educational level, after controlling for beginning salary, is statistically significant, $r_{ab.c}(471) = .28$, $p < .001$, however, it is substantially lower than the bivariate correlation between these variables, $r_s(472) = .66$, $p < .001$. This suggests that much of the correlation between current salary and educational level is simply due to beginning salary. However, there also appears to be a relationship between current salary and educational level that is independent of beginning salary.

Try On Your Own

Try to compute the partial correlation between beginning salary and current salary, after controlling for educational level. Interpret the results and then report them using APA style.

---

[15] For those of you who are interested, degrees of freedom for partial correlation are equal to $N − 2$ – the number of variables being partialled out (and yes, that does mean that you can partial out more than one variable).

4

Regression

Learning Objectives

In this chapter, you will learn how to conduct simple and multiple regression analyses. You will learn how to determine whether a regression model and the individual predictors contained in the model are statistically significant. You will learn how to interpret the multiple correlation, coefficient of determination, and standard error of estimate, as well as partial, semi-partial, and squared semi-partial correlation coefficients. Finally, you will learn how to use the results of a regression analysis to construct the least-squares regression line and make predictions.

We will once again use the sample data file 'Employee data' for the demonstrations in this chapter. If you saved the file used for the demonstrations in Chapter 3, you should use it (because we will once again consider the numerically recoded gender variable). If you didn't save the file, the original 'Employee data' file can be found in your SPSS program files (refer to the section Opening Sample Data Files on page 34 if you forget how to locate and open these files).

SIMPLE REGRESSION

Regression is simply an extension of correlation. We use correlation when we want to determine the strength and direction of relationship between variables, and we use regression when we want to use the information about those relationships to make predictions. We will begin with a simple regression analysis, which is a regression analysis that involves the use of only one predictor variable.

Let's start by examining whether we can use number of years of education to predict beginning salary. Thus, for this analysis beginning salary will be our 'criterion variable' (the variable we want to predict, or the Y variable) and years of education will be our 'predictor variable' (the variable we will use to make our predictions, or the X variable).

Conducting a Simple Regression Analysis
(Analyze→Regression→Linear)

To conduct the analysis, use the upper toolbar to go to **Analyze→Regression→Linear**.

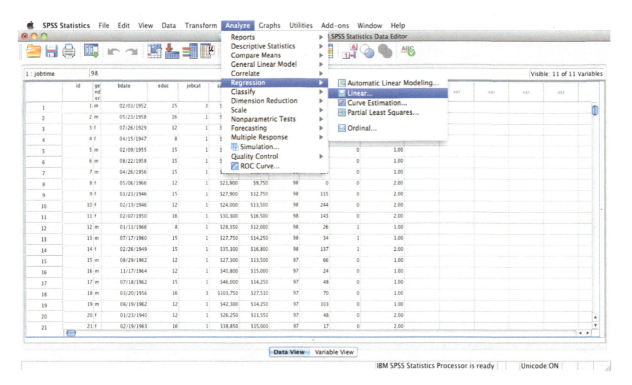

A 'Linear Regression' dialog window like the one shown on the following page will open. Enter **Beginning Salary** (the criterion variable) into the **Dependent box** using the corresponding **blue arrow**. Next, enter **Educational Level** (the predictor variable) into the **Independent(s) box** using the corresponding **blue arrow**. Click **OK**.

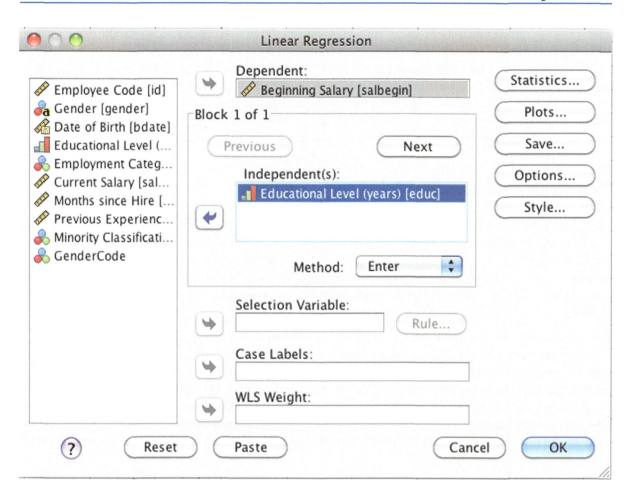

Assessing the Model's Accuracy

A number of tables will now appear in the output window. We will start by reviewing the 'Model Summary' table shown below.

Model Summary

Model	R	R Square	Adjusted R Square	Std. Error of the Estimate
1	.633[a]	.401	.400	$6,098.259

a. Predictors: (Constant), Educational Level (years)

The Correlation Coefficient (r)

The column labeled 'R'[16] displays the size of the correlation between the predictor and criterion variables (r_{XY}). In this case it reflects the size of the correlation between years of education and beginning salary. In the context of regression, this value can also be interpreted as the correlation between actual Y scores and

---

[16] Since we are using simple rather than multiple regression (i.e., we only have one predictor variable), the capitalized R shown in the table should really be a lowercase *r*, but sometimes SPSS doesn't make these subtle distinctions.

what we would predict them to be using the regression equation ($r_{YY'}$). Therefore, we could also interpret this value as the size of the correlation between people's actual beginning salaries and what we would predict their salaries to be using the regression equation. In line with this alternative interpretation, the value displayed in this table will always be positive. As such the direction of relationship between the predictor and criterion cannot be inferred using this value. Note that the higher the correlation coefficient, the more accurate our predictions will be. A correlation of .63 is quite large, so we can expect our predictions of people's beginning salaries based on their years of education to be quite accurate.

The Coefficient of Determination (r^2)

The column labeled 'R Square' contains the value of the coefficient of determination (r^2). In the context of simple regression we use a lowercase r^2 to symbolize the coefficient of determination, but in the context of multiple regression we use an uppercase R^2. The coefficient of determination is an indicator of the proportion of variability of Y that can be accounted for by X. In the context of regression, it is often interpreted as the proportion of variability of Y that can be predicted by X. It is considered an indicator of the effect size.

The table displays a value of .40 for the coefficient of determination, which we would report as $r^2 = .40$. To transform this proportion to a percentage you can simply multiply the value by 100. However, you will first need to click repeatedly on the value in the table to reveal the full decimal remainder in order to keep the value accurate to two decimal places. Based on the revealed value (of .400937), we can now conclude that 40.09% of the variability of Y (beginning salary) can be accounted for and predicted by X (educational level). In other words, about 60% of the variability in beginning salary is still not accounted for and cannot be predicted by educational level (i.e., we don't know what that remaining variability in beginning salary is associated with).[17]

The table also displays an 'Adjusted R Square' value. For simple regression, the adjusted r^2 will always be pretty much identical to r^2. You will notice bigger differences between the adjusted R^2 and R^2 as you add more predictor variables in multiple regression analyses. While R^2 will almost always increase as you add more predictor variables (as long as the predictor variables are correlated with the criterion variable), adjusted R^2 adjusts for the number of predictors used, and its value only increases if a predictor improves the prediction more than would be expected based on chance.

The Standard Error of Estimate (SEE)

The standard error of estimate, which is abbreviated SEE, is an indicator of the average amount of error our predictions will contain. As such, this value allows us to determine how much confidence we should have in our predictions. Specifically, the standard error of estimate reflects the average distance between the actual Y scores in the dataset and what they would be predicted to be using the regression equation.

The value of the standard error of estimate can be found in the column labeled 'Std. Error of the Estimate.' A quick glance at this column reveals that SEE = $6,098.26. This value means that if we use the regression equation to predict people's beginning salaries based on their level of education, then, on average, we can expect our predictions to be off by $6,098.26.

---

[17] Another interpretation of r^2 is that it reflects the proportion reduction in error from using the least-squares regression line to make predictions rather than the mean. In other words, if we use the least-squares regression line to make predictions, our predictions will have 40.09% less error on average than they would if we just predicted the mean of the Y variable for everyone (we predicted everyone would have the average beginning salary).

Assessing Statistical Significance

The Regression Model

The determination of whether the regression model is statistically significant involves using an F statistic to assess whether the coefficient of determination (r^2) is significantly different than 0. In other words, it involves assessing whether we can predict a significant proportion of variability in Y (the criterion) using X (the predictor). If a regression model is statistically significant, it suggests that we can reliability predict the criterion using the predictor. Information on whether r^2, and therefore the regression model, is statistically significant is contained in the 'ANOVA' table shown below.

ANOVA[a]

Model		Sum of Squares	df	Mean Square	F	Sig.
1	Regression	1.175E+10	1	1.175E+10	315.897	.000[b]
	Residual	1.755E+10	472	37188762.8		
	Total	2.930E+10	473			

a. Dependent Variable: Beginning Salary

b. Predictors: (Constant), Educational Level (years)

The F statistic is used to determine the significance of regression models. As shown in the column labeled 'F' the value of the F statistic for this regression model is 315.90. The p value provided under the column labeled 'Sig.' is less than .001. Since the p value is less than .05, this regression model would be considered statistically significant. The table also shows the values of the degrees of freedom, in the column labeled 'df.' F statistics are always associated with two degrees of freedom values. The first degrees of freedom value that we will need to report is listed in the row labeled 'Regression' and the second degrees of freedom value that we will need to report is listed in the row labeled 'Residual.' These values are reported in parentheses, separated by a column. So, they would be reported as: (1, 472).

Since the assessment of the significance of the regression model involves using the F statistic to evaluate whether the coefficient of determination is statistically significant, it is customary to report the coefficient of determination along with the F statistic, degrees of freedom, and p value. To illustrate, using APA style, we would report that the regression model is statistically significant, $r^2 = .40$, $F(1, 472) = 315.90$, $p < .001$.

The Predictor

In addition to determining whether the regression model is significant, it is typically important to assess whether the individual predictors are statistically significant. Since simple regression involves only one predictor variable, if the overall regression model is significant, then the predictor will be always be significant, and vice versa. As you will see, this is not always the case for multiple regression, which involves more than one predictor variable. The determination of whether the predictor is statistically significant involves using a t statistic to assess whether the standardized slope (β) is significantly different than 0. If a predictor is statistically significant, it suggests that we can use it to reliability predict the criterion.

Information on whether β, and therefore the predictor variable, is statistically significant is contained in the 'Coefficients' table shown on the following page. Specifically, you will find information about the significance of the predictor in the row labeled with the name of the predictor, in this case the row labeled 'Educational Level (years).' The value of the slope (β) is provided in the column labeled 'Beta' and the value of the t statistic is displayed in the column labeled 't'. As shown in the table, the value of β is .63 and the value of t is 17.77. The adjacent column shows that the p value is less than .001. Once again, since this p value is less than .05, it would typically be considered statistically significant. The degrees of freedom that need to be reported along with the t statistic can be found in the 'ANOVA' table shown previously, in the row labeled 'Residual.' As shown in that table, the degrees of freedom are equal to 472. Using APA style, we would report these statistics in the following manner: β = .63, $t(472)$ = 17.77, p < .001. Note that β is a Greek character and therefore should not be italicized.

Coefficients[a]

Model		Unstandardized Coefficients		Standardized Coefficients	t	Sig.
		B	Std. Error	Beta		
1	(Constant)	-6290.967	1340.920		-4.692	.000
	Educational Level (years)	1727.528	97.197	.633	17.773	.000

a. Dependent Variable: Beginning Salary

Constructing the Equation for the Least-Squares Regression Line

The 'Coefficients' table shown above also contains the regression coefficients, which we need to construct the equation for the least-squares regression line. The equation for the least-squares regression line is: $Y' = bX + a$. Alternatively, this formula may be expressed as: $\hat{Y} = a + bX$.

Y' or $\hat{Y}$ represents the criterion variable. Remember, the criterion variable is the variable we want to predict. The symbol b represents the unstandardized slope of the least-squares regression line. It indicates the amount we would predict the criterion (Y) to increase for every one-unit increase in the predictor variable (X). The X in the equation represents the person's score on the predictor variable. We will later substitute in values for X in order to make our predictions. Finally, the symbol a represents the intercept of the least-squares regression line. It is the place where the least-squares regression line intersects with the y-axis. It can also be interpreted as the value of Y we would predict for someone with a score of 0 on the predictor variable (X).

The table above displays the values of the regression coefficients (slope and intercept) in the column labeled 'B.' Note that although SPSS uses a capitalized B, the unstandardized slope is typically symbolized with a lower case b (this is true for simple and multiple regression). The slope is presented in the row labeled with the predictor variable—in this case 'Educational Level (years).' The table displays the value of the slope of the least-squares regression line as 1727.53. This value means that for every one-year increase in education, we would predict an increase in beginning salary of $1,727.53. The intercept (a) is presented in the row labeled '(Constant).' The table shows that the intercept is −6290.97. This value indicates that we would predict a salary of −$6,290.97 for a person with 0 years of education (apparently we would predict that the person is going into debt!). We now have all of the information we need to construct the equation of the least-squares regression line. We simply need to substitute these values of b and a into the equation: $Y' = bX + a$

$$Y' = 1727.53X - 6290.97$$

Once again, the value listed in the column labeled 'Beta' provides the standardized slope. The standardized slope reflects the slope of the least-squares regression line for the standardized (z transformed) variables. You can make predictions of people's Y scores in standardized units (z score units) simply by multiplying their z scores on the X variable by this beta (β) value. That is: $z_{Y'} = βz_X$. Importantly, this formula can only be used for standardized variables; only z scores on the X variable can be inputted and only z scores on the

Y variable can be predicted. For simple regression, the value of beta will always be the same as the value of the correlation between the predictor and criterion.

Using the Regression Equation to Make Predictions

Now that we have constructed the equation for the least-squares regression line, we can make predictions of people's beginning salaries based on their years of education. What beginning salary would we predict for an individual with 16 years of education? To answer this question, all we need to do is substitute 16 (the person's score on the *X* variable) in for *X* and solve for *Y'*. Let's go ahead and do that:

$$Y' = 1727.5283X - 6290.9673$$

$$Y' = 1727.5283(16) - 6290.9673$$

$$Y' = 21349.49$$

Thus, we would predict a beginning salary of $21,349.49 for a person with 16 years of education. Apparently these data are quite outdated; let's hope you'll be earning more than that with your four years of postsecondary education! Note that we used values rounded to four decimal places when we were using the regression equation to make a prediction in order to increase the accuracy of our prediction and we only rounded the final predicted value to two decimal places. While the SPSS output only displays three decimal places, the remaining decimal places can be revealed by repeatedly clicking on the value in the table in the output window.

Reporting the Results

There is no single way to report the results of a regression analysis. With that said, the coefficient of determination, the *F* statistic, the slope, and the *t* statistic should all be reported along with their respective degrees of freedom and *p* values. Note that the coefficient of determination should be reported along with the *F* statistic since it is this value that is evaluated in determining the statistical significance of a model. Similarly, the slope is evaluated in determining the significance of the predictor, so the beta weight or unstandardized slope of the predictor should be reported along with the *t* statistic. Typically, if the results are being published to address a theoretical question, the beta weight (β) is reported, but if the results are being published for the purpose of using the regression model to make predictions, then the unstandardized slope (*b*) is reported. Additional statistics may be reported depending on their importance to the specific research question under investigation. Using APA style, we could report the following:

> Educational level was found to be a significant predictor of beginning salary, $\beta = .63$, $t(472) = 17.77$, $p < .001$. Moreover, educational level was found to account for 40.09% of the variability in beginning salary, $F(1, 472) = 315.90$, $p < .001$.

MULTIPLE REGRESSION WITH TWO PREDICTOR VARIABLES

In this section, you will learn how to perform a multiple regression analysis. Multiple regression is simply an extension of simple regression. Simple regression is used to make predictions using only one predictor variable, while multiple regression is used to make predictions using more than one predictor variable.

We will begin by considering the simplest case of multiple regression, multiple regression with two predictor variables. Let's say we want to predict people's beginning salaries based on their years of education *and* their previous work experience. Now we are in the land of multiple regression because we are using more than one predictor variable. In this example, beginning salary is the criterion variable (i.e., the variable we want to predict) and it is labeled Y', just as it was in simple regression. Educational level and previous experience are the predictor variables (i.e., the variables we will use to make our predictions). The predictor variables are labeled X, as in simple regression, but since there are now multiple predictor variables we need to distinguish them from each other using subscripts. Thus, we will label educational level X_1 and previous experience X_2.

Conducting a Multiple Regression Analysis
(Analyze→Regression→Linear)

To run the analysis, go to **Analyze→Regression→Linear**. A 'Linear Regression' dialog window like the one shown below will open. Move **Beginning Salary** (the criterion variable) into the **Dependent box** using the corresponding **blue arrow**. Next, enter **Educational Level** and **Previous Experience** (the predictor variables) into the **Independent(s) box** using the corresponding **blue arrow**.

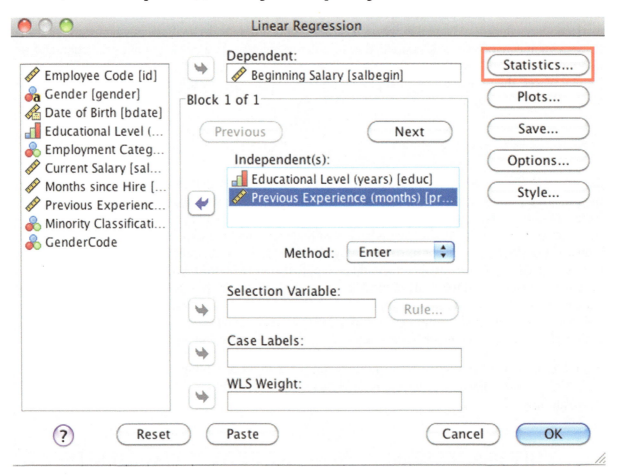

We will also consider the partial and semi-partial correlations for the predictor variables since they are commonly used statistics in the context of multiple regression. To obtain these statistics you need to click on the **Statistics button** highlighted in the image shown above. This will open the 'Linear Regression: Statistics' dialog window shown on the next page.

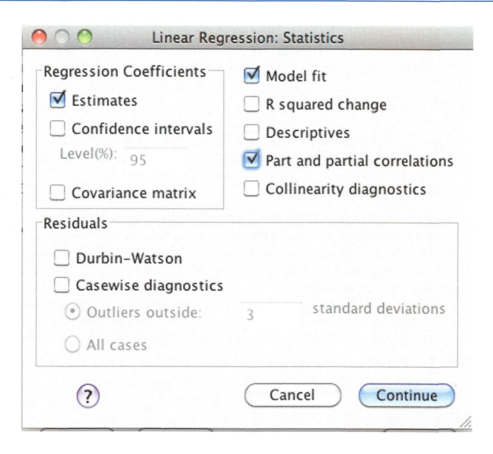

Check the option for **Part and partial correlations**. Click **Continue** and then **OK** to close the dialog windows and execute the analysis.

Assessing the Model's Accuracy

We will begin by examining the 'Model Summary' table shown below.

Model Summary

Model	R	R Square	Adjusted R Square	Std. Error of the Estimate
1	.668[a]	.446	.443	$5,871.763

a. Predictors: (Constant), Previous Experience (months), Educational Level (years)

The Multiple Correlation (R)

The multiple correlation is displayed in the column labeled 'R.' Now that we have multiple predictors (i.e., multiple X variables), it is appropriate to use a capitalized R to symbolize this correlation. The multiple correlation is the correlation between the criterion and the best linear combination of the set of predictor variables. The multiple correlation can also be interpreted as the correlation between the actual Y scores (the scores in the dataset) and what they would be predicted to be using the regression equation (Y').

As such a higher multiple correlation, indicates more accurate predictions. As you can see above, the value of the multiple correlation is .67. This value is quite high, suggesting that our predictions of people's beginning salaries using educational level and previous experience will be quite accurate.

The Coefficient of Multiple Determination (R^2)

The coefficient of multiple determination (R^2) is an indicator of the proportion of variability of Y that can be accounted for by the best linear combination of the *set* of predictors. Once again, in the context of regression it is often interpreted as the proportion of variability of Y that can be predicted by the set of predictors. It is considered an indicator of the effect size.

The value of the coefficient of multiple determination is provided in the column labeled 'R Square.' As shown in the table, $R^2 = .45$. If you click on that value repeatedly you will see the value rounded to four decimal places is .4458. Based on this value we can conclude that 44.58% of the variability in beginning salary can be predicted (or accounted for) by years of education and months of previous experience. That means that about 55.42% of the variability in beginning salary is still not accounted for and cannot be predicted using these two variables. If you'll recall, we were able to account for 40.09% of the variability in beginning salary when we were using only educational level as a predictor variable. By adding previous experience as a second predictor variable, we are able to account for more of the variability in beginning salary (an additional 4.49%).

You will notice that now that we have more than one predictor, the difference between the R^2 value and the adjusted R^2 value is slightly larger. However, the difference of .003 is still trivial.

The Standard Error of Estimate (SEE)

The standard error of estimate, displayed in the column labeled 'Std. Error of the Estimate,' is $5,871.76. This means that if we use the regression equation to make predictions of people's beginning salaries, we can expect our predictions to be off by $5,871.76, on average. You should note that adding previous experience as a second predictor variable reduced the error in our predictions from $6,098.26 (as shown in the 'Model Summary' table in the Simple Regression section) to $5,871.76. It appears that our prediction accuracy has been improved by adding the second predictor variable, previous experience.

Assessing Statistical Significance

The Regression Model

In order to determine whether the regression model is statistically significant (to determine whether we can predict a significant proportion of variability in beginning salary and therefore whether we can use the model to reliably make predictions), we need to consult the 'ANOVA' table shown below.

ANOVA[a]

Model		Sum of Squares	df	Mean Square	F	Sig.
1	Regression	1.306E+10	2	6.531E+9	189.427	.000[b]
	Residual	1.624E+10	471	34477599.6		
	Total	2.930E+10	473			

a. Dependent Variable: Beginning Salary

b. Predictors: (Constant), Previous Experience (months), Educational Level (years)

The table clearly shows that the regression model is statistically significant. The value of the F statistic is 189.43, the p value is less than .001, and the degrees of freedom are 2 (see row labeled 'Regression') and 471 (see row labeled 'Residual'). Therefore, $R^2 = .45$, $F(2, 471) = 189.43$, $p < .001$.

The Predictors

Next, we need to consider the significance of the individual predictors. The 'Coefficients' table shown below provides this information. The row labeled 'Educational Level (years)' provides information about the predictor variable educational level, while the row labeled 'Previous Experience (months)' provides information about the predictor variable previous experience. The values of the standardized slopes (β) and t statistics can be found in the columns labeled 'Beta' and 't' respectively, while the p values can be found in the column labeled 'Sig.' The table shows that both predictors have p values less than .001, and therefore both would be considered statistically significant. The degrees of freedom for the t statistic can once again be found in the ANOVA table shown previously, in the row labeled 'Residual.' As shown in that table, the value of the degrees of freedom is equal to 471.

Bringing it all together, we can now report that years of education, $\beta = .69$, $t(471) = 19.42$, $p < .001$, and months of previous experience, $\beta = .22$, $t(471) = 6.17$, $p < .001$, are statistically significant predictors of beginning salary.

Coefficients^a

Model		Unstandardized Coefficients		Standardized Coefficients	t	Sig.	Correlations		
		B	Std. Error	Beta			Zero-order	Partial	Part
1	(Constant)	−9902.786	1417.474		−6.986	.000			
	Educational Level (years)	1878.211	96.717	.688	19.420	.000	.633	.667	.666
	Previous Experience (months)	16.470	2.668	.219	6.174	.000	.045	.274	.212

a. Dependent Variable: Beginning Salary

Constructing the Equation for the Least-Squares Regression Line

We can also use the information in the 'Coefficients' table shown above to construct the equation for the least-squares regression line. The multiple regression equation for two predictor variables is: $Y' = b_1X_1 + b_2X_2 + a$. Alternatively, this formula may be expressed as: $\hat{Y} = a + bX_1 + bX_2$. The equations are very similar to the simple regression equations; they have simply been expanded to include two predictor variables (X_1 and X_2) and each of their slopes (b_1 and b_2).

The values of the intercept (a) and unstandardized slopes (b_1 and b_2) are presented in the column labeled 'B.' The slope for each predictor variable is presented in the row labeled with its name. The table shows that b_1 (the slope for educational level) is 1878.21, and b_2 (the slope for previous experience) is 16.47. The value of the slope for educational level means that for every one-year increase in education, we would predict an increase in beginning salary of $1,878.21 (assuming months of previous experience is held constant). Similarly, the value of the slope for previous experience means that for every one-month increase in previous experience, we would predict an increase in beginning salary of $16.47 (assuming years of education is held constant). The intercept is once again presented in the row labeled '(Constant)'; its value is −9902.79. This value indicates that we would predict a beginning salary of −$9,902.79 for a person with 0 years of education and 0 months of previous experience. We now have all of the information we need to construct the equation of the least-squares regression line. We simply need to substitute these values in for b_1, b_2, and a into the equation: $Y' = b_1X_1 + b_2X_2 + a$

$$Y' = 1878.21X_1 + 16.47X_2 - 9902.79$$

The values listed in the column labeled 'Beta' represent the standardized regression coefficients. Again, they are symbolized β, and they reflect the slope of the least-squares regression line for the standardized (z transformed) variables. While the regression coefficients provided in the column labeled 'B' are unstandardized, the beta values provided in the column labeled 'Beta' are standardized. Since the beta values are standardized, you can directly compare their values in order to determine which variable will be given the most weight in making the predictions. For this reason, these beta values are often referred to as beta weights. Since the beta weight associated with educational level (.69) is higher than the beta weight associated with previous experience (.22), we can determine that people's educational level will be weighted more heavily in making predications of their beginning salaries than their previous experience.

You should note that it is not possible to directly compare the unstandardized values of the slopes (the b values) since their units of measure will typically differ from one another (indeed, in this case educational level is measured in the unit of years, while previous experience is measured in the unit of months). Comparing unstandardized b values is like comparing apples with oranges.

Using the Regression Equation to Make Predictions

Since both of our predictor variables are statistically significant, we can begin making predictions of people's beginning salaries based on their years of education and months of previous experience. What beginning salary would we predict for an individual with 16 years of education and 110 months of previous experience? To answer this question all we need to do is substitute 16 in for X_1 and 110 in for X_2 and then solve for Y'. Let's go ahead and do that:

$$Y' = 1878.2114(X_1) + 16.4704(X_2) - 9902.7861$$

$$Y' = 1878.2114(16) + 16.4704(110) - 9902.7861$$

$$Y' = 21960.34$$

Thus, we would predict a beginning salary of $21,960.34 for a person with 16 years of education and 110 months of previous experience. Once again, please note that we used values rounded to four decimal places when we were using the regression equation to make a prediction in order to increase the accuracy of our prediction, and we only rounded our final answer to two decimal places.

Bivariate, Partial, Semi-Partial, and Squared Semi-Partial Correlations

Since we selected the option to compute the part and partial correlations prior to executing the analysis, the 'Coefficients' table will also include a section labeled 'Correlations' (highlighted in the image shown below). This section includes the values of the bivariate, partial, and semi-partial correlations between each of the predictors and the criterion.

Coefficients[a]

Model		Unstandardized Coefficients		Standardized Coefficients	t	Sig.	Correlations		
		B	Std. Error	Beta			Zero-order	Partial	Part
1	(Constant)	-9902.786	1417.474		-6.986	.000			
	Educational Level (years)	1878.211	96.717	.688	19.420	.000	.633	.667	.666
	Previous Experience (months)	16.470	2.668	.219	6.174	.000	.045	.274	.212

a. Dependent Variable: Beginning Salary

Bivariate Correlations (r)

The column labeled 'Zero-order' displays the bivariate correlations, which are simply the correlations between each of the predictors and the criterion. As established previously, the correlation between educational level and beginning salary is equal to .63. The table also shows that the correlation between previous experience and beginning salary is equal to .05.

Partial Correlations ($r_{ab.c}$)

As described in Chapter 3, a partial correlation is a correlation between two variables after statistically controlling for a potential third variable. The column labeled 'Partial' provides the partial correlations between each of the predictors and the criterion, after statistically controlling for the other predictor variable. More precisely, the partial correlation for X_1 can be interpreted as the relationship between X_1 and Y after the influence of X_2 on both X_1 and Y has been removed. So, it is the relationship between the predictor and criterion after the influence of the other predictor on both the predictor of interest and the criterion has been removed.

As you can see, the partial correlation for educational level is .67. This means that the correlation between educational level and beginning salary, after the influence of previous experience on both educational level and beginning salary has been removed, is .67. Since there is little difference between the bivariate and partial correlations, we can determine that previous experience has little to no influence on the relationship between educational level and beginning salary.

The partial correlation for previous experience is .27. This means that the correlation between previous experience and beginning salary, after the influence of educational level on both previous experience and beginning salary has been removed, is .27. You may be surprised to see that controlling for the influence of years of education increased the size of the correlation between previous experience and beginning salary. When a partial (or semi-partial) correlation is higher than the original bivariate correlation, it indicates the presence of a suppression effect. Suppression is an advanced statistical concept and suppression effects can be quite difficult to interpret. Basically, partial correlations involve removing variability in the predictor and criterion that is accounted for (or predicted) by the other predictor(s). When we see a suppression effect like this, it suggests that by controlling for the other predictor variable(s) we are removing some error variance (variability in Y and/or X that are unrelated to each other), and in doing so we are making the variable a better predictor.

Semi-Partial Correlations ($r_{a(b.c)}$)

SPSS refers to semi-partial correlations by their old-school name, part correlation. The semi-partial correlation is a close cousin of the partial correlation. It also provides an indicator of the relationship between two variables after statistically controlling for a potential third variable, however the degree of control differs somewhat. For *partial* correlation, the influence of the potential third variable on *both* the criterion and the other predictor variable is statistically controlled. In contrast, for *semi-partial correlation*, the influence of the potential third variable on *only* the other predictor variable is controlled. Precisely, the semi-partial correlation for X_1 can be interpreted as the relationship between X_1 and Y after the influence of X_2 on X_1 has been removed. The symbol for the semi-partial correlation is $r_{a(b.c)}$. It may help you to remember that since the parentheses are around only the b and c, only the influence of c (X_2) on b (X_1) is being controlled.

The semi-partial correlation for educational level of .67 displayed in the table indicates that the correlation between educational level and beginning salary, after the influence of previous experience on educational level has been removed, is .67. The fact that the bivariate and semi-partial correlations for educational level do not differ much from one another suggests that controlling for the influence of previous experience on educational level does little to the magnitude of the correlation between educational level and beginning salary. Finally, the semi-partial correlation for previous experience indicates that the correlation between previous experience and beginning salary, after the influence of educational level on previous experience has been removed, is .21. Once again, since the semi-partial correlation is higher than the bivariate correlation we know that a suppression effect is at play here.

Squared Semi-Partial Correlations ($r^2_{a(b.c)}$)

The squared semi-partial correlation, denoted $r^2_{a(b.c)}$, is a very informative statistic in the context of multiple regression. Its value can be interpreted as the proportion of variability in the criterion that can be uniquely predicted by the predictor. In other words, it is an indicator of the proportion of the criterion (the variable we are trying to predict) that only that specific predictor variable can predict. As such, this statistic provides an indicator of the unique contribution of a predictor variable to a regression model. More specifically, the squared semi-partial correlation tells us how much R^2 will decrease if that predictor variable is removed from the regression model (or how much R^2 has increased as a result of including the predictor).

The squared semi-partial correlation is simply the semi-partial correlation squared. Thus, the squared semi-partial correlation for the predictor variable educational level is .44 ($.6661^2 = .4437 = .44$). This value indicates that educational level uniquely predicts 44.37% of the variability of beginning salary, and that if the variable educational level was removed from the regression model the value of R^2 would decrease by .44. The squared semi-partial correlation for the predictor variable previous experience is .04 ($.2118^2 = .0449 = .04$). This value indicates that previous experience uniquely predicts about 4.49% of the variability of beginning salary, and that if it was removed from the regression model the value of R^2 would decrease by only .04. While this decrease seems trivial, the fact that previous experience is a significant predictor of beginning salary suggests that this decrease would be statistically significant.

Reporting the Results

We could report these results in the following manner:

> Educational level and previous experience were found to account for a significant proportion of variability in beginning salary, $R^2 = .45$, $F(2, 471) = 189.43$, $p < .001$. Moreover, both educational level, $\beta = .69$, $t(471) = 19.42$, $p < .001$, and previous experience, $\beta = .22$, $t(471) = 6.17$, $p < .001$, were found to be significant predictors of beginning salary.

MULTIPLE REGRESSION WITH THREE PREDICTOR VARIABLES

To illustrate multiple regression with three predictor variables (one of which is a nominal variable with two categories), we will try to predict beginning salary using educational level, previous experience, and gender. We will label educational level X_1, previous experience X_2, and gender X_3.

Since gender was originally entered as a string (text) variable, we will not be able to perform any analyses with it until it has been recoded using a numeric code. If you are not using the data file you used in Chapter 3, that contains the numerically recoded gender variable (GenderCode), you will need to recode the gender variable. To recode the gender variable, go to **Transform→Recode into Different Variables**. Name the recoded variable **GenderCode** and then proceed to recode the current gender code **m** to **1** and **f** to **2** (refer to the section Recoding Variables on pages 16–18 if you forget how to do this).

Conducting a Multiple Regression Analysis
(Analyze→Regression→Linear)

To conduct the multiple regression analysis you will need to go to **Analyze→Regression→Linear**. Using the 'Linear Regression' dialog window shown below, move **Beginning Salary** (the criterion variable) into the **Dependent box** using the corresponding **blue arrow**. Next, move **Educational Level**, **Previous Experience**, and **GenderCode** (all three predictor variables) into the **Independent(s) box** using the corresponding **blue arrow**.

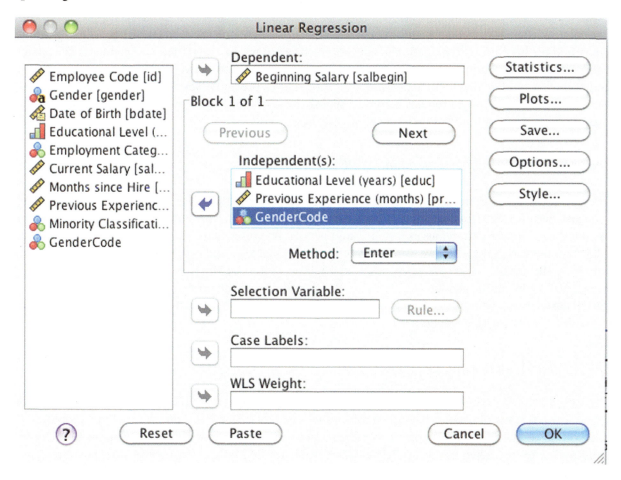

Next, click on the **Statistics button** to open the 'Linear Regression: Statistics' dialog widow. Check the option for **Part and partial correlations** displayed in that dialog window. Click **Continue** and then **OK** to close both dialog windows.

Assessing the Model's Accuracy

We will begin with the 'Model Summary' table shown on the following page.

Model Summary

Model	R	R Square	Adjusted R Square	Std. Error of the Estimate
1	.696[a]	.484	.481	$5,671.155

a. Predictors: (Constant), GenderCode, Previous Experience (months), Educational Level (years)

The Multiple Correlation (R)

The column labeled 'R' shows the multiple correlation is .70. Therefore, the correlation between beginning salary and the best linear combination of educational level, previous experience, and gender is .70. Alternatively, this value can be interpreted to indicate that the correlation between actual beginning salaries and what we would predict them to be using educational level, previous experience, and gender is .70. The value of the multiple correlation has increased slightly by adding the third predictor variable, suggesting that the accuracy of our prediction has been slightly improved.

The Coefficient of Multiple Determination (R^2)

The 'R Square' value in the table shows that the coefficient of multiple determination is .48. This value indicates that 48.41% of the variability in beginning salary is accounted for (and can be predicted by) years of education, months of previous experience, and gender. As you may recall, we were able to account for 44.58% of the variability when we were using only educational level and previous experience as predictor variables. This means we can predict an additional 3.83% (48.41 – 44.58 = 3.83) of the variability in beginning salary by including gender as a predictor. If you understood the previous section on squared semi-partial correlations, you should already be able to determine that the squared semi-partial correlation for the coded gender variable will be .0383.

The Standard Error of Estimate (SEE)

The table provides a value of $5,671.16 for the standard error of estimate. This means that we can expect that our predictions of people's beginning salaries will be off by $5,671.16, on average. You should note that adding gender as a third predictor variable reduced the average amount of error in our predictions from $5,871.76 (as shown in the 'Model Summary' table in the Multiple Regression with Two Predictor Variables section) to $5,671.16. This provides us with even more evidence that our prediction accuracy has been improved by adding the third predictor variable.

Assessing Statistical Significance

The Regression Model

In order to determine whether the regression model is statistically significant (to determine whether we can reliably predict beginning salary using the set of three predictors) we need to consult the 'ANOVA' table shown on the following page.

ANOVA[a]

Model		Sum of Squares	df	Mean Square	F	Sig.
1	Regression	1.418E+10	3	4.728E+9	147.014	.000[b]
	Residual	1.512E+10	470	32162000.3		
	Total	2.930E+10	473			

a. Dependent Variable: Beginning Salary

b. Predictors: (Constant), GenderCode, Previous Experience (months), Educational Level (years)

The table clearly shows that the regression model is statistically significant. As shown in the table, the value of the F statistic is 147.01, the p value is less than .001, and the degrees of freedom are 3 and 470. Once again, this would be reported as $R^2 = .48$, $F(3, 470) = 147.01$, $p < .001$.

The Predictors

The 'Coefficients' table shown below displays p values less than .001 for all three predictor variables, and therefore all three would be considered statistically significant. Recall that the degrees of freedom can be found in the 'ANOVA' table displayed above in the row labeled 'Residual.' As shown in that table, the degrees of freedom are equal to 470. Using this information and the information provided in the columns labeled 'Beta' and 't,' we can now report that the predictors educational level, $\beta = .60$, $t(470) = 15.82$, $p < .001$, previous experience, $\beta = .16$, $t(470) = 4.47$, $p < .001$, and gender, $\beta = -.22$, $t(470) = -5.91$, $p < .001$, are all statistically significant.

Coefficients[a]

Model		Unstandardized Coefficients		Standardized Coefficients	t	Sig.	Correlations		
		B	Std. Error	Beta			Zero-order	Partial	Part
1	(Constant)	-1045.042	2030.198		-.515	.607			
	Educational Level (years)	1625.292	102.753	.596	15.817	.000	.633	.589	.524
	Previous Experience (months)	12.001	2.685	.159	4.469	.000	.045	.202	.148
	GenderCode	-3446.504	583.307	-.218	-5.909	.000	-.457	-.263	-.196

a. Dependent Variable: Beginning Salary

Constructing the Equation for the Least-Squares Regression Line

The 'Coefficients' table shown above also contains all of the values we need to construct the equation for the least-squares regression line. The equation used for a model with three predictor variables is: $Y' = b_1X_1 + b_2X_2 + b_3X_3 + a$. Alternatively, this formula may be expressed as: $\hat{Y} = a + bX_1 + bX_2 + bX_3$. This equation is very similar to the equation for multiple regression with two predictor variables; it has simply been expanded to include a third predictor variable (X_3) and its slope (b_3).

Let's begin by finding the values of the regression coefficients (slopes and intercept). Once again, they are all presented in the column labeled 'B.' The table shows that b_1 (the slope for educational level) is 1625.29, b_2 (the slope for previous experience) is 12.00, and b_3 (the slope for gender) is −3446.50.

The negative value of b_3 indicates an inverse slope. In other words, the negative value of b_3 means that the relationship between the predictor and criterion is negative (i.e., that as the value of the predictor variable increases, the value of the criterion variable decreases). Since gender is a nominal variable and we used numeric codes to represent the values of this variable, we need to carefully consider the codes that we used before we can understand the nature of this relationship. We used a code of 1 to represent the male gender and a code of 2 to represent the female gender. As such, the negative slope associated with this variable indicates that being female (a higher value) is associated with lower beginning salaries, and similarly, that being male (a lower value) is associated with higher beginning salaries. By considering the value of the slope, we can determine that being female is associated with a decrease in the predicted beginning salary of $3,446.50 (assuming educational level and previous experience are held constant).

The intercept, shown in the row labeled '(Constant),' is −1045.04. We now have all of the information we need to construct the equation of the least-squares regression line. All we need to do is substitute in the values of b_1, b_2, b_3, and a into the equation: $Y' = b_1X_1 + b_2X_2 + b_3X_3 + a$

$$Y' = 1625.29X_1 + 12.00X_2 - 3446.50X_3 - 1045.04$$

By examining the standardized beta weights listed in the column labeled 'Beta,' we can see that educational level is still the strongest predictor (it has the highest beta weight and therefore will be weighted most in making our predictions). Gender has the second highest beta weight and will therefore be weighted less than educational level but more than previous experience in making our predictions. Finally, previous experience has the lowest beta weight and will therefore be given the least weight in our predictions.

Using the Regression Equation to Make Predictions

Since all of our predictors are statistically significant, we can now use the regression equation to predict people's beginning salaries based on their years of education, months of previous experience, and gender. Let's say that Joe (a male) has 16 years of education and 80 months of previous experience. What would we predict his beginning salary to be? To answer this question, we simply need to substitute these values in for the corresponding Xs in our regression equation and then solve the equation. Since we labeled educational level X_1, previous experience X_2, and gender X_3, we need to substitute 16 for X_1, 80 for X_2, and 1 (the numeric code we have assigned for males) for X_3. Let's go ahead and do that:

$$Y' = 1625.2918X_1 + 12.0014X_2 - 3446.5036X_3 - 1045.0423$$

$$Y' = 1625.2918(16) + 12.0014(80) - 3446.5036(1) - 1045.0423$$

$$Y' = 22473.23$$

Thus, we would predict Joe's beginning salary to be $22,473.23. Try to predict what Jane's (a female's) beginning salary would be if she had the same number of years of education and same months of previous experience as Joe. You will discover that her beginning salary would be predicted to be $3,446.50 less than Joe's salary (which is the value of b_3).

Bivariate, Partial, Semi-Partial, and Squared Semi-Partial Correlations

The 'Correlations' section of the 'Coefficients' table (highlighted below) also shows the bivariate, partial, and semi-partial correlations for all three predictor variables.

Coefficients[a]

Model		Unstandardized Coefficients B	Std. Error	Standardized Coefficients Beta	t	Sig.	Correlations Zero-order	Partial	Part
1	(Constant)	−1045.042	2030.198		−.515	.607			
	Educational Level (years)	1625.292	102.753	.596	15.817	.000	.633	.589	.524
	Previous Experience (months)	12.001	2.685	.159	4.469	.000	.045	.202	.148
	GenderCode	−3446.504	583.307	−.218	−5.909	.000	−.457	−.263	−.196

a. Dependent Variable: Beginning Salary

Bivariate Correlations (r)

The column labeled 'Zero-order' shows that the bivariate correlation between educational level and beginning salary is .63, the correlation between previous experience and beginning salary is .05, and the correlation between gender and beginning salary is −.46. Once again, since we coded males with 1s (a lower value) and females with 2s (a higher value), this negative correlation indicates that being female is associated with earning a lower beginning salary.

Partial Correlations ($r_{ab.c}$)

The column labeled 'Partial' shows the partial correlations for all three predictor variables. In the case of a model with three predictors, the partial correlation for X_1 can be interpreted as the relationship between X_1 and Y after the influences of X_2 and X_3 on both X_1 and Y have been removed. So, it is the relationship between the predictor and criterion after the influences of the other predictor variables on both the predictor of interest and the criterion have been removed.

The partial correlation of .59 for educational level provided in the table indicates that the correlation between educational level and beginning salary, after the influences of previous experience and gender on both educational level and beginning salary have been removed, is. 59. The partial correlation of .20 displayed for previous experience indicates that the correlation between previous experience and beginning salary, after the influences of educational level and gender on both previous experience and beginning salary have been removed, is .20. Finally, the partial correlation of −.26 for the gender variable shown above indicates that the correlation between gender and beginning salary, after the influences of educational level and previous experience on both gender and beginning salary have been removed, is −.26.

Semi-Partial Correlations ($r_{a(b.c)}$)

The column labeled 'Part' shows the semi-partial correlations for all three predictor variables. In the case of a regression model with three predictors, the semi-partial correlation for X_1 can be interpreted as the relationship between X_1 and Y after the influences of X_2 and X_3 on X_1 have been removed. So, it is the relationship between the predictor and criterion after the influences of the other predictors on the predictor of interest have been removed.

The semi-partial correlation of .52 for educational level displayed above indicates that the correlation between educational level and beginning salary, after the influences of previous experience and gender on educational level have been removed, is .52. The semi-partial correlation of .15 displayed for previous experience indicates that the correlation between previous experience and beginning salary, after the influences of educational level and gender on previous experience have been removed, is .15. Finally, the semi-partial correlation of −.20 provided for the gender variable informs us that the correlation between gender and beginning salary, after the influences of educational level and previous experience on gender have been removed, is −.20.

Squared Semi-Partial Correlations ($r^2_{a(b.c)}$)

To review, the squared semi-partial correlation indicates the proportion of variability in the criterion that the predictor (and only that predictor) can predict. In other words, it is an indicator of the unique contribution of a predictor variable to a regression model, and therefore its value tells us how much R^2

will decrease if that predictor variable is removed from the regression model (similarly, it informs us how much R^2 has increased as a result of the inclusion of the predictor in the model).

The squared semi-partial correlation for the predictor variable educational level is .27 ($.5240^2 = .2746$ = .27). This value indicates that educational level uniquely predicts 27.46% of the variability of beginning salary (i.e., 27.46% of the variability in beginning salary is being predicted by educational level alone). Therefore, if the variable educational level was removed from the regression model, the value of R^2 would decrease by .27. The squared semi-partial correlation for the predictor variable previous experience is .02 ($.1481^2 = .0219 = .02$). This value indicates that previous experience uniquely predicts 2.19% of the variability of beginning salary, and that if the variable previous experience was removed from the regression model, the value of R^2 would decrease by only .02. Finally, the squared semi-partial correlation for the gender variable is .04 ($-.1958^2 = .0383 = .04$). This value indicates that of all of the variability in beginning salary, 3.83% of it is being predicted uniquely by gender.[18]

Reporting the Results

An example of an APA style write-up of these results is:

> The three predictors were found to account for just under half of the variability in beginning salary ($R^2 = .48$), which was statistically significant, $F(3, 470) = 147.01$, $p < .001$. Educational level, $\beta = .60$, $t(470) = 15.82$, $p < .001$, previous experience, $\beta = .16$, $t(470) = 4.47$, $p < .001$, and gender, $\beta = -.22$, $t(470) = -5.91$, $p < .001$, were all found to be significant predictors of beginning salary, with women receiving significantly lower beginning salaries than men. An examination of the squared semi-partial correlations revealed that educational level uniquely predicts 27.46% of the variability in beginning salary, 3.83% of the variance in beginning salary can be uniquely predicted by gender, and previous experience uniquely accounts for only 2.19% of the variance in beginning salaries.

MULTIPLE REGRESSION WITH FOUR PREDICTOR VARIABLES

To illustrate multiple regression with four predictor variables (one of which is not a significant predictor), we will try to predict beginning salary using educational level, previous experience, gender, and months since hire. We will label educational level X_1, previous experience X_2, gender X_3, and months since hire X_4.

Conducting a Multiple Regression Analysis
(Analyze→Regression→Linear)

To conduct the analysis you will need to go to **Analyze→Regression→Linear**. Using the 'Linear Regression' dialog window put **Beginning Salary** (the criterion variable) into the **Dependent box**. Next, move **Educational Level**, **Previous Experience**, **GenderCode**, and **Months since Hire** (all four predictor variables) into the **Independent(s) box**. Once again, click on the **Statistics button** to open the 'Linear Regression: Statistics' dialog widow. Check the option for **Part and partial correlations** displayed in that dialog window. Click **Continue** and then **OK** to close both dialog windows and execute the analysis.

---

[18] You may notice that the sum of all of the squared semi-partial correlations does not equal the value of R^2. This is because some of the value of R^2 is made up of shared, rather than unique, variance (it reflects variability of Y that can be predicted by more than one of the predictor variables).

Assessing the Model's Accuracy

Model Summary

Model	R	R Square	Adjusted R Square	Std. Error of the Estimate
1	.699[a]	.488	.484	$5,655.231

a. Predictors: (Constant), Months since Hire, Previous Experience (months), GenderCode, Educational Level (years)

The Multiple Correlation (R)

By referring to the column labeled 'R' in the 'Model Summary' table shown above, you should be able to see that the multiple correlation is .70. There has been almost no increase in the value of the multiple correlation by adding months since hire as a predictor variable. This suggests that the accuracy of our prediction will not be improved much from the previous model we considered.

The Coefficient of Multiple Determination (R^2)

The coefficient of multiple determination shown in the table is .49. By repeatedly clicking on this value in the output table you should be able to determine that 48.81% of the variability in beginning salary is accounted for (and can be predicted by) the set of four predictor variables. As you may recall, we were able to account for 48.41% of the variability when we were using the set of three predictors. Therefore, adding months since hire as a fourth predictor doesn't really allow us to account for much more of the variability in beginning salary. Specifically, we can only predict an additional 0.40% of the variability in beginning salary by including months since hire as a predictor.

The Standard Error of Estimate (SEE)

The 'Model Summary' table also displays a value of $5,655.23 for the standard error of estimate. Therefore, we can expect that our predictions of people's beginning salaries will be off by about $5,655.23, on average. Once again, you should note that adding the fourth predictor variable reduced the average amount of error in our predictions from $5,671.16 (as shown in the 'Model Summary' table in the Multiple Regression with Three Predictor Variables section above) to $5,655.23, which is a trivial amount.

Assessing Statistical Significance
The Regression Model

Next, we will consider the 'ANOVA' table shown below to determine whether we can reliably predict people's beginning salaries using the set of four predictors.

ANOVA[a]

Model		Sum of Squares	df	Mean Square	F	Sig.
1	Regression	1.430E+10	4	3.575E+9	111.795	.000[b]
	Residual	1.500E+10	469	31981636.4		
	Total	2.930E+10	473			

a. Dependent Variable: Beginning Salary

b. Predictors: (Constant), Months since Hire, Previous Experience (months), GenderCode, Educational Level (years)

The table clearly shows that the regression model is statistically significant. The value of the F statistic is 111.79[19], the p value is less than .001, and the degrees of freedom that we will need to report are 4 and 469. Once again, this would be reported as $R^2 = .49$, $F(4, 469) = 111.79$, $p < .001$.

The Predictors

Next, we need to consider the significance of the individual predictors. As you can see in the 'Coefficients' table shown below, the three predictors we considered previously have p values less than .001. However, the p value for the variable months since hire is greater than .05, and therefore this predictor is not statistically significant. Recall that the degrees of freedom can be found in the 'ANOVA' table in the row labeled 'Residual.' As shown in that table, the degrees of freedom are equal to 469. Using this information and the information provided in the columns labeled 'Beta' and 't' we can now report that the predictors educational level, $\beta = .60$, $t(469) = 15.90$, $p < .001$, previous experience, $\beta = .16$, $t(469) = 4.48$, $p < .001$, and gender, $\beta = -.22$, $t(469) = -6.01$, $p < .001$, are statistically significant. However, months since hire is not a statistically significant predictor of beginning salary, $\beta = -.06$, $t(469) = -1.91$, $p = .06$.

Coefficients[a]

Model		Unstandardized Coefficients		Standardized Coefficients	t	Sig.	Correlations		
		B	Std. Error	Beta			Zero-order	Partial	Part
1	(Constant)	2988.860	2925.067		1.022	.307			
	Educational Level (years)	1630.014	102.495	.597	15.903	.000	.633	.592	.525
	Previous Experience (months)	12.004	2.678	.160	4.483	.000	.045	.203	.148
	GenderCode	-3503.090	582.423	-.222	-6.015	.000	-.457	-.268	-.199
	Months since Hire	-49.507	25.911	-.063	-1.911	.057	-.020	-.088	-.063

a. Dependent Variable: Beginning Salary

Constructing the Equation for the Least-Squares Regression Line

The equation for the least-squares regression line that is used for four predictor variables is: $Y' = b_1X_1 + b_2X_2 + b_3X_3 + b_4X_4 + a$. Alternatively, this formula may be expressed as: $\hat{Y} = a + bX_1 + bX_2 + bX_3 + bX_4$.

The column labeled 'B' lists the values of the regression coefficients (slopes and intercept). By looking at the table, you should be able to see that b_1 (the slope for educational level) is 1630.01, b_2 (the slope

[19] Note that the value in the second decimal remainder has been rounded down because the actual value of the F statistic (found by repeatedly clicking on the value in the table) is 111.794761, which rounds to 111.79.

for previous experience) is 12.00, b_3 (the slope for the gender variable) is −3503.09, and b_4 (the slope for months since hire) is −49.51.

Once again, the negative values of b_3 and b_4 indicate that our coded gender variable and the months since hire variable have inverse slopes (that as the value of one of these predictor variables increases, the value of the criterion variable decreases). It appears that more months since hire are associated with lower beginning salaries. So, people who were hired longer ago were offered lower beginning salaries. It makes sense that over time beginning salaries would rise (with inflation rates), however, the predictor is not significant, so the relationship may just reflect chance variation rather than a real effect.

The intercept, shown in the row labeled '(Constant),' is −2988.86. We now have all of the information we need to construct the equation of our regression line. All we need to do is substitute in the values of b_1, b_2, b_3, and a into the equation: $Y' = b_1X_1 + b_2X_2 + b_3X_3 + b_4X_4 + a$

$$Y' = 1630.01X_1 + 12.00X_2 - 3503.09X_3 - 49.51X_4 - 2988.86$$

By examining the standardized beta weights in the column labeled 'Beta,' we can see that educational level is still the strongest predictor (it has the highest beta weight and therefore will be weighted most in making our predictions), gender is the second strongest predictor, previous experience is the third strongest predictor, and months since hire is the weakest predictor of beginning salary.

Using the Regression Equation to Make Predictions

Since one of the predictor variables is not statistically significant, we should not use this equation for the least-squares regression line to make predictions of people's beginning salaries. As the months since hire variable is not statistically significant, we should revert back to using our previous three-predictor regression equation to make predictions.

Bivariate, Partial, Semi-Partial, and Squared Semi-Partial Correlations

Finally, we will examine the 'Correlations' section of the 'Coefficients' table (highlighted below).

Coefficients[a]

Model		Unstandardized Coefficients		Standardized Coefficients	t	Sig.	Correlations		
		B	Std. Error	Beta			Zero-order	Partial	Part
1	(Constant)	2988.860	2925.067		1.022	.307			
	Educational Level (years)	1630.014	102.495	.597	15.903	.000	.633	.592	.525
	Previous Experience (months)	12.004	2.678	.160	4.483	.000	.045	.203	.148
	GenderCode	-3503.090	582.423	-.222	-6.015	.000	-.457	-.268	-.199
	Months since Hire	-49.507	25.911	-.063	-1.911	.057	-.020	-.088	-.063

a. Dependent Variable: Beginning Salary

Bivariate Correlations (r)

The column labeled 'Zero-order' shows that the bivariate correlation between educational level and beginning salary is .63, the correlation between previous experience and beginning salary is .05, the correlation between gender and beginning salary is −.46, and the correlation between months since hire and beginning salary is −.02.

You should note that the bivariate correlations are not affected by the presence of the other predictor variables (so the first three values are identical to the ones reported previously when we considered the regression model with three predictor variables). However, this is not the case for the partial and semi-partial correlations, which change whenever a predictor is added or removed from the regression model.

Partial Correlations ($r_{ab.c}$)

The column labeled 'Partial' shows the partial correlations for all four predictor variables. In the case of a model with four predictors, the partial correlation for X_1 can be interpreted as the relationship between X_1 and Y after the influences of X_2, X_3, and X_4 on both X_1 and Y have been removed. So, it is the relationship between the predictor and criterion after the influences of the other predictor variables on both the predictor of interest and the criterion have been removed.

The partial correlation of .59 for the educational level variable displayed in the table indicates that the correlation between educational level and beginning salary, after the influences of previous experience, gender, and months since hire on both educational level and beginning salary have been removed, is .59. The partial correlation of .20 for the previous experience variable shown above indicates that the correlation between previous experience and beginning salary, after the influences of educational level, gender, and months since hire on both previous experience and beginning salary have been removed, is .20.

Try On Your Own

You should try to interpret the remaining two partial correlations on your own for practice.

Semi-Partial Correlations ($r_{a(b.c)}$)

The column labeled 'Part' shows the semi-partial correlations for all four predictor variables. In the case of a regression model with four predictors, the semi-partial correlation for X_1 can be interpreted as the relationship between X_1 and Y after the influences of X_2, X_3, and X_4 on X_1 have been removed. Once again, it is the relationship between the predictor and criterion after the influences of the other predictors on the predictor of interest have been removed.

The semi-partial correlation of .53 for the educational level variable shown above indicates that the correlation between educational level and beginning salary, after the influences of previous experience, gender, and months since hire on educational level have been removed, is .53. The semi-partial correlation of .15 for the previous experience variable shown above indicates that the correlation between previous experience and beginning salary, after the influences of educational level, gender, and months since hire on previous experience have been removed, is .15.

Try On Your Own

You should try to interpret the remaining two semi-partial correlations on your own for practice.

Squared Semi-Partial Correlations ($r^2_{a(b.c)}$)

The squared semi-partial correlation indicates the proportion of variability in the criterion that the predictor (and that predictor alone) can predict. In other words, it is an indicator of the unique contribution of a predictor variable to a regression model, and therefore, its value tells us how much R^2 will decrease if that predictor variable is removed from the regression model.

The squared semi-partial correlation for the predictor variable educational level is .28 ($.5254^2 = .2760 = .28$). This value indicates that educational level uniquely predicts 27.60% of the variability of beginning salary (i.e., 27.60% of the variability in beginning salary is being predicted by educational level alone). Therefore, if the variable educational level was removed from the regression model, the value of R^2 would decrease by .28. The squared semi-partial correlation for the predictor variable previous experience is .02 ($.1481^2 = .0219 = .02$). This value indicates that previous experience uniquely predicts 2.19% of the variability of beginning salary, and that if this variable was removed from the regression model, the value of R^2 would decrease by only .02.

Try On Your Own

You should try to calculate and interpret the remaining two squared semi-partial correlations on your own for practice.

Reporting the Results

A sample APA style write-up of these results is:

Just under half of the variability of beginning salary was predicted by the set of four predictors ($R^2 = .49$). While this was found to be statistically significant, $F(4, 469) = 111.79$, $p < .001$, only three of the predictors showed a significant relationship with the criterion. Specifically, years of education, $\beta = .60$, $t(469) = 15.90$, $p < .001$, months of previous experience, $\beta = .16$, $t(469) = 4.48$, $p < .001$, and gender, $\beta = -.22$, $t(469) = -6.01$, $p < .001$, were all found to be significant predictors of beginning salary, with women receiving significantly lower beginning salaries than men. In contrast, number of months since hire was not a significant predictor of beginning salary, $\beta = -.06$, $t(469) = -1.91$, $p = .06$, and could uniquely predict only 0.40% of the variability in these salaries.

Try On Your Own

The least-squares regression equation can be expanded to include five, six, or more predictors in the same way that it was expanded to include two, three, and four predictors (by adding in additional b and X terms). Try to build and use an equation with five predictors as additional practice and write up the results using APA style. Try using educational level, previous experience, gender, minority status, and current salary to predict beginning salary.

5

Advanced Regression

Learning Objectives

In this chapter, you will learn how to conduct hierarchical and stepwise regression analyses. You will also learn how to interpret the results of these analyses and report them using APA style.

Now that you have mastered simple and multiple regression, you are ready for an introduction to more advanced forms of regression. We will once again use the sample data file 'Employee data' for the demonstrations in this chapter. If you saved the file used for the demonstrations in Chapters 3 and 4, you should use it because we will once again consider the numerically recoded gender variable.

HIERARCHICAL REGRESSION

Hierarchical regression is simply an extension of multiple regression. It allows us to directly compare two regression models (one simpler and one more complex) in order to determine whether the addition of a predictor (or a set of predictors) makes a significant contribution to the regression model. It is commonly used to determine whether a predictor can predict a significant amount of variability in the criterion, *over and above* another predictor (or set of predictors). Similarly, it is commonly used to determine whether a predictor variable is significantly related to a criterion after statistically controlling for one or more other predictor variables.

Imagine that you are an industrial organizational consultant who has been hired to assist a group of women in making a case against their company for offering lower beginning salaries to the female employees. The company is countering the women's argument by stating that years of education is the biggest factor in determining the beginning salary offered to each employee, and the reason women tend to earn lower beginning salaries is that they tend to have fewer years of education. In line with their counter-argument, you find a significant relationship between gender and educational level. In order to assist these women, you will need to examine whether the gender discrepancy in beginning salaries can be accounted for by differences in educational level. To explore this possibility, you will need to examine whether adding gender to a regression model predicting beginning salary significantly improves the prediction accuracy over and above (i.e., after controlling for) educational level. Thus, for this analysis beginning salary will be the criterion variable (the Y variable), and years of education and gender will be the predictor variables (the X variables).

Conducting a Hierarchical Regression Analysis
(Analyze→Regression→Linear)

To conduct the hierarchical regression analysis, go to **Analyze→Regression→Linear**. A 'Linear Regression' dialog window like the one shown below will open. Put **Beginning Salary** (the criterion variable) into the

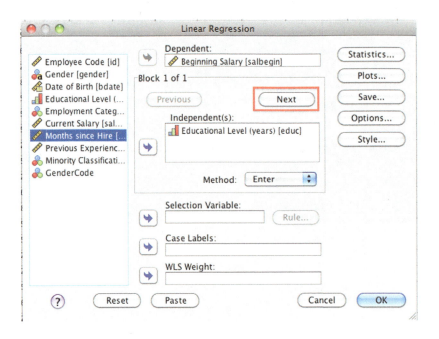

Dependent box using the corresponding **blue arrow**. Next, move **Educational Level** (the variable you want to control for) into the **Independent(s) box** using the corresponding **blue arrow**.

Now, in order to make this a hierarchical regression analysis you simply need to click on the **Next button** (highlighted in the image shown on the preceding page), and then put **GenderCode** (your primary predictor variable of interest) into the blank **Independent(s) box** that will appear.

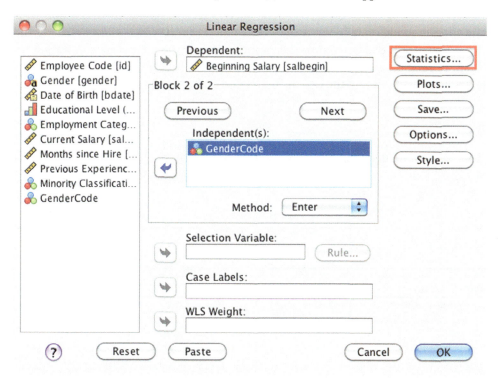

Next, click on the **Statistics button** (highlighted in the image shown above). This will open the 'Linear Regression: Statistics' dialog window shown below. Check the option for **R squared change**.

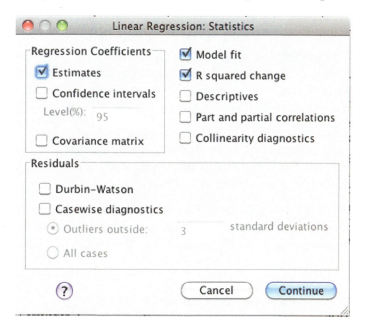

Click on **Continue** and then **OK** to close the dialog windows and execute the analysis.

Interpreting the Results

As stated previously, hierarchical regression allows us to directly compare two regression models (one simpler and one more complex) in order to determine whether the addition of a predictor (or a set of predictors) contributes significantly to the regression model. As such, each of the results tables will include the results of two regression models: Model 1 and Model 2.

Model 1 is the simpler model that contains only the predictor(s) inputted before clicking the next button in the Linear Regression dialog window (i.e., it includes only the predictors you want to control for). Model 2 is the more complex model that contains all of the predictor variables (those inputted both before and after clicking the next button in the Linear Regression dialog window).

Model Comparisons

We will begin by examining the 'Model Summary' table shown below as it contains the primary results of interest.

Model Summary

Model	R	R Square	Adjusted R Square	Std. Error of the Estimate	Change Statistics				
					R Square Change	F Change	df1	df2	Sig. F Change
1	.633[a]	.401	.400	$6,098.259	.401	315.897	1	472	.000
2	.680[b]	.462	.460	$5,784.256	.061	53.637	1	471	.000

a. Predictors: (Constant), Educational Level (years)

b. Predictors: (Constant), Educational Level (years), GenderCode

The results shown in the row labeled '1' are for a simple regression analysis using only educational level to predict beginning salary. As such, they are redundant with those computed for the simple regression analysis reviewed in Chapter 4. The column labeled 'R' shows that the correlation between educational level and beginning salary is .63, the column labeled 'R Square' shows that the coefficient of determination is .40, and the column labeled 'Std. Error of the Estimate' indicates that the standard error of estimate is $6,098.26.

Information about the significance of Model 1 can be found in both the 'ANOVA' table shown on the following page and in the 'Change Statistics' section of the 'Model Summary' table shown above. Despite the name of this section, since for the first model there is no change to consider, the values displayed in the top section pertain to the significance of the model as a whole.[20] By reviewing these statistics, you should be able to determine that the first model is statistically significant, $r^2 = .40$, $F(1, 472) = 315.90$, $p < .001$.

The results displayed in the row labeled '2' are for the second, more complex model that includes both educational level and gender as predictors. First, by reviewing the left side of the table you should be able to find that the value of the multiple correlation is .68, the value of the coefficient of multiple determination is .46, and the standard error of estimate is $5,784.26.

For the second, more complex model, the label 'Change Statistics' is appropriate because the statistics displayed on the right side of the 'Model Summary' table concern the change in the R^2 value that resulted from adding the predictor variable gender. The column labeled 'R Square Change' shows that adding the

[20]You can confirm this for yourself by comparing the F, df, and p values provided in the top section of the 'Model Summary' table with those provided in the upper portion of the 'ANOVA' table, and by comparing the R Square and R Square Change statistics provided in the 'Model Summary' table.

variable gender increased the R^2 value by .06, and the column labeled 'F Change' shows that the F statistic for this change in R^2 is equal to 53.64. Finally, you should be able to see that this change is statistically significant because the associated p value, displayed in the last column, is less than .001. Change statistics are symbolized using Δ. Therefore, these results would be reported as: $\Delta R^2 = .06$, $\Delta F(1, 471) = 53.64$, $p < .001$. These results indicate that there is a relationship between gender and beginning salary that is independent of educational level. It looks like these women do have a case against this company after all, because differences in years of education do not account for the gender discrepancy in beginning salary; women are paid significantly lower beginning salaries even after years of education are taken into consideration.

Model Statistics

The 'ANOVA' table shown below simply provides the degrees of freedom, F statistics, and p values for both of the regression models. As you can once again see, the simpler model that includes only educational level as a predictor is statistically significant, $r^2 = .40$, $F(1, 472) = 315.90$, $p < .001$. As you should expect then, the more complex model that includes both educational level and gender as predictors is also statistically significant, $R^2 = .46$, $F(2, 471) = 202.38$, $p < .001$. However, these results are not terribly interesting and in most cases we would not report them. For hierarchical regression, we focus more on the change statistics described above.

ANOVA[a]

Model		Sum of Squares	df	Mean Square	F	Sig.
1	Regression	1.175E+10	1	1.175E+10	315.897	.000[b]
	Residual	1.755E+10	472	37188762.8		
	Total	2.930E+10	473			
2	Regression	1.354E+10	2	6.771E+9	202.381	.000[c]
	Residual	1.576E+10	471	33457613.3		
	Total	2.930E+10	473			

a. Dependent Variable: Beginning Salary

b. Predictors: (Constant), Educational Level (years)

c. Predictors: (Constant), Educational Level (years), GenderCode

Predictor Statistics

Finally, the 'Coefficients' table shown on the following page displays the regression coefficients (in both unstandardized and standardized form), the t statistics, and the p values for each of the predictors, separately for each of the regression models. Once again, the statistics for the simpler (one predictor) model are presented in the upper portion of the table (the portion labeled '1'). You should be able to see that educational level is a significant predictor of beginning salary, $\beta = .63$, $t(472) = 17.77$, $p < .001$. The value of the degrees of freedom that need to be reported with the t statistic can be found in the 'Model Summary' table displayed on page 84 in the column labeled 'df2' and the row labeled '1.'

The statistics for the more complex (two predictor) model are presented in the lower portion of the 'Coefficients' table (the portion labeled '2'). By referring to this portion of the table, you should be able to see that both educational level, $\beta = .54$, $t(471) = 14.90$, $p < .001$, and gender, $\beta = -.26$, $t(471) = -7.32$, $p < .001$, are significant predictors. Moreover, the slope of the gender variable is inverse, confirming that being female is associated with being paid lower beginning salaries. Recall that the degrees of freedom can be found in the 'Model Summary' table in the column labeled 'df2' and the row labeled '2.'

Coefficients[a]

Model		Unstandardized Coefficients		Standardized Coefficients	t	Sig.
		B	Std. Error	Beta		
1	(Constant)	−6290.967	1340.920		−4.692	.000
	Educational Level (years)	1727.528	97.197	.633	17.773	.000
2	(Constant)	3265.086	1822.140		1.792	.074
	Educational Level (years)	1470.321	98.655	.539	14.904	.000
	GenderCode	−4180.769	570.853	−.265	−7.324	.000

a. Dependent Variable: Beginning Salary

Reporting the Results

We could report the following results to the company:

> A hierarchical regression analysis was conducted using gender to predict beginning salary after entering years of education. The results showed that the inclusion of gender significantly improved the prediction accuracy of the model, $\Delta R^2 = .06$, $\Delta F(1, 471) = 53.64$, $p < .001$. Moreover, gender was a significant predictor of beginning salary, $\beta = -.26$, $t(471) = -7.32$, $p < .001$, even after statistically controlling for years of education. These results clearly show that there is a relationship between gender and beginning salary that is independent of educational level. In other words, women are paid significantly lower beginning salaries even after years of education are taken into consideration.

Conducting a More Complex Hierarchical Regression Analysis
(Analyze→Regression→Linear)

In the previous example, our simpler model contained only one predictor variable and our more complex model contained only two predictor variables. But hierarchical regression can be used to examine models that are far more complex.

Imagine that the company under investigation disputed the results you presented to them and claimed that women also have fewer months of previous experience, which, combined with fewer years of education, accounts for the gender discrepancy in beginning salary. At the same time, a group of minorities heard about your case and, based on their findings that minorities also earn significantly lower beginning salaries, they asked to join the case against this company. Thus, for this more complex hierarchical regression analysis you will need to examine whether gender and minority status significantly predict beginning salary after controlling for educational level and previous experience. This will allow you to determine whether lower levels of education and fewer months of previous experience account for the lower beginning salaries offered to women and minorities.

To run the analysis, go to **Analyze→Regression→Linear**. A 'Linear Regression' dialog window like the one shown on the next page will open. Put **Beginning Salary** (the criterion variable) into the **Dependent box** using the corresponding **blue arrow**. Next, move **Educational Level** and **Previous Experience** (the variables you want to control for) into the **Independent(s) box** using the corresponding **blue arrow**.

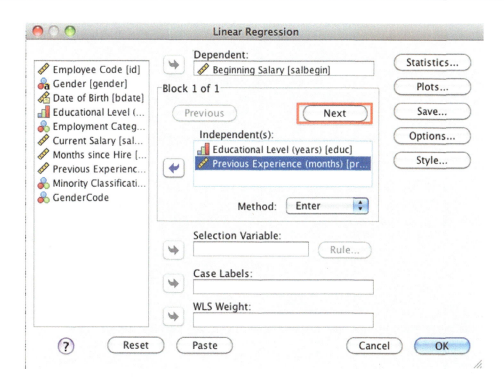

Click on the **Next button** (highlighted in the image shown above), and then put **GenderCode** and **Minority Classification** (your primary predictor variables of interest) into the blank **Independent(s) box** that will appear in the dialog window.

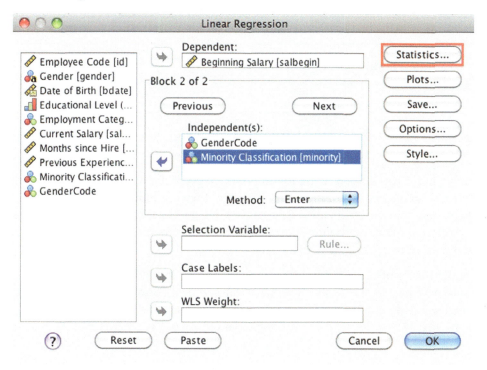

Next, click on the **Statistics button** (highlighted in the image shown above). This will open the 'Linear Regression: Statistics' dialog window shown on the next page. Check the option for **R squared change**.

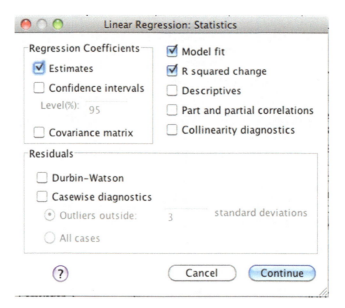

Click on **Continue** and then **OK** to close the dialog windows and execute the analysis.

Interpreting the Results

Model Comparisons

Once again, the 'Model Summary' table shown below contains the primary results of interest.

Model Summary

Model	R	R Square	Adjusted R Square	Std. Error of the Estimate	R Square Change	F Change	df1	df2	Sig. F Change
					Change Statistics				
1	.668ᵃ	.446	.443	$5,871.763	.446	189.427	2	471	.000
2	.706ᵇ	.499	.494	$5,596.641	.053	24.723	2	469	.000

a. Predictors: (Constant), Previous Experience (months), Educational Level (years)

b. Predictors: (Constant), Previous Experience (months), Educational Level (years), Minority Classification, GenderCode

The results shown in the row labeled '1' are for a multiple regression analysis using educational level and previous experience to predict beginning salary. As such, they are redundant with those computed for the two-predictor multiple regression analysis reviewed in Chapter 4. Consistent with the results of that analysis, the results show that the multiple correlation is .67, the coefficient of multiple determination is .45, and the standard error of estimate is $5,871.76.

Information about the significance of Model 1 can once again be found in both the 'ANOVA' table shown on the following page and in the 'Model Summary' table shown above, in the section labeled 'Change Statistics.' By reviewing these statistics, you should be able to determine that the model is statistically significant, $R^2 = .45$, $F(2, 471) = 189.43$, $p < .001$. Once again, since this is the first model, there is no change to consider.

The results displayed in the row labeled '2' are for the second, more complex model that includes educational level, previous experience, gender, and minority classification as predictors. First, by reviewing the left side of the table you should be able to see that for this more complex model, the value of the multiple correlation is .71, the value of the coefficient of multiple determination is .50, and the standard error of estimate is $5,596.64.

The right side of the 'Model Summary' table shows that adding gender and minority classification to the regression model increased the R^2 value by .05, and the F statistic for this change in R^2 is equal to 24.72. Finally, you should be able to see that this change is statistically significant because the associated p value is less than .001. These results would be reported as $\Delta R^2 = .05$, $\Delta F(2, 469) = 24.72$, $p < .001$.

These results indicate that, as a set, gender and minority classification account for a significant proportion of variability in beginning salary, over and above that accounted for by educational level and previous experience. In other words, gender and minority classification together show a significant relationship with beginning salary after education and previous experience are statistically controlled. It appears that fewer years of education and fewer months of previous experience do not account for the diminished salaries offered to these groups. Our case against this company has been strengthened.

Predictor Statistics

Since we added more than one predictor variable to the second step of the hierarchical regression analysis, we need to examine the significance of each of the predictors before we can reach too strong of a conclusion. Let's skip over the 'ANOVA' table for a minute and jump right to the 'Coefficients' table shown below.

Coefficients[a]

Model		Unstandardized Coefficients		Standardized Coefficients	t	Sig.
		B	Std. Error	Beta		
1	(Constant)	-9902.786	1417.474		-6.986	.000
	Educational Level (years)	1878.211	96.717	.688	19.420	.000
	Previous Experience (months)	16.470	2.668	.219	6.174	.000
2	(Constant)	405.585	2041.776		.199	.843
	Educational Level (years)	1574.258	102.343	.577	15.382	.000
	Previous Experience (months)	12.812	2.659	.170	4.818	.000
	GenderCode	-3670.725	578.845	-.233	-6.341	.000
	Minority Classification	-2339.717	634.479	-.123	-3.688	.000

a. Dependent Variable: Beginning Salary

Once again, the statistics for the simpler (two-predictor) model are presented in the upper portion of the table. Consistent with the two-predictor multiple regression analysis we conducted in Chapter 4, the results show that educational level, $\beta = .69$, $t(471) = 19.42$, $p < .001$, and previous experience, $\beta = .22$, $t(471) = 6.17$, $p < .001$, are each significant predictors of beginning salary. Note that the degrees of freedom that need to be reported can be found in the 'Model Summary' table shown on page 88 in the column labeled 'df2' and the row labeled '1.'

The statistics for the more complex (four-predictor) model are presented in the lower portion of the 'Coefficients' table. As displayed in this portion of the table, both gender, $\beta = -.23$, $t(469) = -6.34$, $p < .001$, and minority classification, $\beta = -.12$, $t(469) = -3.69$, $p < .001$, are significant predictors of beginning salary. The degrees of freedom that need to be reported can be found in the 'Model Summary' table shown on page 88 in the column labeled 'df2' and the row labeled '2.' Once again, the slope for each of these predictors is inverse, confirming that being female (coded with a higher value) is associated with lower beginning salaries, and being a minority (also coded with a higher value) is associated with lower beginning salaries.

Moreover, by directly comparing the beta weights for gender ($\beta = -.23$) and minority classification ($\beta = -.12$), we can determine that after controlling for educational level and previous experience, the relationship between gender and beginning salary is stronger than the relationship between minority status and beginning salary. By examining the unstandardized regression coefficients, we can determine that we would predict a woman in this company to be paid a beginning salary that is $3,670.73 lower than the beginning salary of a man with the same number of years of education, months of previous experience, and minority classification. Similarly, we would predict a minority to be paid a beginning salary that is $2,339.72 lower than a non-minority of the same gender, with the same number of years of education, and the same number of months of previous experience. Apparently, this company is both sexist and racist, but they are a bit more sexist than racist.

Note that if one of the predictors showed a significant bivariate correlation with the criterion but was no longer a significant predictor when entered into the second step of the hierarchical regression model, it would suggest that the predictors entered into the first step of the model could explain the relationship between the non-significant predictor and the criterion. Assume we found that the R^2 change and F change statistics were statistically significant (that adding the set of predictors significantly improved the prediction accuracy of the model), that the predictor gender was statistically significant, but that the predictor minority classification was *not* significant. This pattern of findings would suggest that while years of education and months of previous experience do not account for the gender discrepancy in beginning salary, years of education and previous experience *do* account for the decreased beginning salaries offered to minorities in this company.

Model Statistics

Finally, although we already have all of the information we need, we will briefly consider the 'ANOVA' table displayed below.

ANOVA[a]

Model		Sum of Squares	df	Mean Square	F	Sig.
1	Regression	1.306E+10	2	6.531E+9	189.427	.000[b]
	Residual	1.624E+10	471	34477599.6		
	Total	2.930E+10	473			
2	Regression	1.335E+10	3	4.450E+9	131.141	.000[c]
	Residual	1.595E+10	470	33935753.3		
	Total	2.930E+10	473			

a. Dependent Variable: Beginning Salary

b. Predictors: (Constant), Previous Experience (months), Educational Level (years)

c. Predictors: (Constant), Previous Experience (months), Educational Level (years), Minority Classification

The 'ANOVA' table simply provides the degrees of freedom, F statistics, and p values for both of the regression models. As you can once again see, the simpler model that includes educational level and previous experience as predictors is statistically significant, $R^2 = .45$, $F(2, 471) = 189.43$, $p < .001$. Accordingly, the more complex model that includes all four predictor variables is also statistically significant, $R^2 = .50$, $F(3, 470) = 131.14$, $p < .001$. Once again, these results are not terribly interesting and in most cases we would not report them.

Reporting the Results

We could report the following results to the company:

A hierarchical regression analysis was conducted using gender and minority classification to predict beginning salary after entering years of education and months of previous experience. The results showed that the inclusion of gender and minority classification significantly improved the prediction accuracy of the regression model, $\Delta R^2 = .05$, $\Delta F(2, 469) = 24.72$, $p < .001$. Moreover, both gender and minority classification continued to be significant predictors of beginning salary after controlling for educational level and previous experience, $\beta = -.23$, $t(469) = -6.34$, $p < .001$, $\beta = -.12$, $t(469) = -3.69$, $p < .001$, respectively. These results clearly show that there are relationships between beginning salary and both gender and minority classification that are independent of educational level and previous experience. In other words, women and minorities are paid significantly lower beginning salaries even after years of education and previous experience are taken into consideration.

STEPWISE REGRESSION

While hierarchical regression is used to address theoretical questions about the nature of relationships between variables, stepwise regression is an entirely exploratory technique. Stepwise regression is empirically, rather than theoretically based. Researchers using stepwise regression enter all variables of interest into the regression model and SPSS finds the model that offers the best predictive utility. It begins by entering the predictor variable that has the strongest relationship to the criterion (i.e., the strongest bivariate correlation with the criterion), and it continues to enter predictor variables, one at a time, in order of their strength of relationship to the criterion (in order of the magnitude of their semi-partial correlations). It stops once the addition of predictors ceases to significantly improve the predictive accuracy of the model (until the addition of more predictors ceases to increase R^2 by a significant amount).

Stepwise regression is not a highly respected technique because as scientists we typically prefer models that are built based on theories. With stepwise regression, a mindless computer program is building the model. Moreover, typically we build models using the data from a sample with the ultimate goal of generalizing to the population. Stepwise regression is often said to capitalize on chance because the decisions about which variables to include in the model are dependent on potentially minor differences in semi-partial correlations derived from a single sample, even though some variability in these statistics from sample to sample is expected. It is often said to "over-fit" the data because the model derived from a single sample is too close to the sample and may not generalize well to the population. For this reason, the models often cannot be successfully replicated with different samples from the same population. We are less likely to capitalize on chance when we use a theory to make decisions about which predictors to include and which models to test. The bottom line is that it is not very respectable to have a mindless computer program making your decisions for you.

Nevertheless, if we are simply interested in exploring relationships between variables and finding the model that offers the best predictive utility, stepwise regression can be a valuable technique. Using the 'Employee data' file with the numerically recoded gender variable we will attempt to find the model that best predicts *current* salary.

Conducting a Stepwise Regression Analysis
(Analyze→Regression→Linear)

To run the analysis, go to **Analyze→Regression→Linear**. A 'Linear Regression' dialog window like the one shown on the following page will open. Make sure that any variables that were entered for the previous analysis are cleared out of the Dependent and Independent(s) boxes (be sure to click the Previous tab to clear out variables entered into the first step of the hierarchical regression analysis we performed previously). Enter **Current Salary** (the criterion variable) into the **Dependent box** using the corresponding

blue arrow. Next, enter **Educational Level, Beginning Salary, Months since Hire, Previous Experience, Minority Classification**, and **GenderCode** into the **Independent(s) box** using the corresponding **blue arrow**. Note that we will not enter Employee Code or Date of Birth because it would not make any sense whatsoever to include either of those variables as predictors (we don't want to be entirely mindless here). Also, we will not enter Employment Category because it is a nominal variable with three levels. It is not appropriate to include nominal variables with more than two levels in any regression analysis (for the same reason these variables should not be used in a correlation analysis). Next, click on the button labeled **Method** (highlighted in the image shown below), and select **Stepwise**.

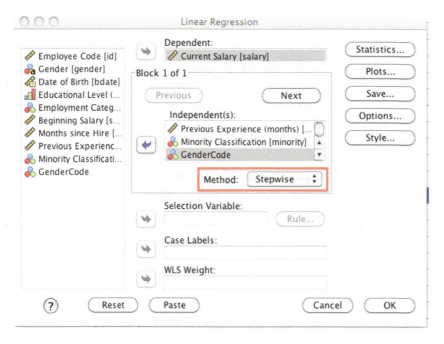

Next, click on the **Statistics button** to open the 'Linear Regression: Statistics' dialog window shown below. Check the option for **R squared change**.

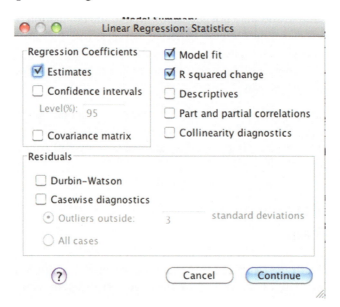

Click **Continue** and then **OK** to close the dialog windows and execute the analysis.

Interpreting the Results

Several tables will now appear in the output window. The first table, labeled 'Variables Entered/Removed,' simply shows the order the variables were entered. This information is also provided in the other tables, so we will not consider it further. Instead, we will begin by examining the 'Model Summary' table shown below.

Model Summary

Model	R	R Square	Adjusted R Square	Std. Error of the Estimate	Change Statistics				
					R Square Change	F Change	df1	df2	Sig. F Change
1	.880[a]	.775	.774	$8,115.356	.775	1622.118	1	472	.000
2	.891[b]	.793	.793	$7,776.652	.019	43.010	1	471	.000
3	.897[c]	.804	.803	$7,586.187	.010	24.948	1	470	.000
4	.900[d]	.810	.809	$7,465.139	.007	16.366	1	469	.000
5	.902[e]	.814	.812	$7,410.457	.003	7.947	1	468	.005

a. Predictors: (Constant), Beginning Salary

b. Predictors: (Constant), Beginning Salary, Previous Experience (months)

c. Predictors: (Constant), Beginning Salary, Previous Experience (months), Months since Hire

d. Predictors: (Constant), Beginning Salary, Previous Experience (months), Months since Hire, Educational Level (years)

e. Predictors: (Constant), Beginning Salary, Previous Experience (months), Months since Hire, Educational Level (years), GenderCode

Finding the Best Model

The 'Model Summary' table shows five of the models SPSS created. The footnote under the table labeled 'a' shows that Model 1 was built using a simple regression analysis that included only beginning salary as a predictor of current salary. The program started with this model because, of all of the predictors we entered, beginning salary shows the highest bivariate correlation with current salary. Since, as shown in the first row, that model was significant, $r^2 = .77$, $F(1, 472) = 1622.12$, $p < .001$, the program proceeded to run a multiple regression analysis.

As shown next to the footnote labeled 'b,' the first multiple regression analysis included beginning salary and previous experience as predictors of current salary. Previous experience was selected next because it shows the highest semi-partial correlation with current salary. As shown in the row labeled '2,' the change in R^2 associated with the addition of this predictor was statistically significant, $\Delta R^2 = .02$, $\Delta F(1, 471) = 43.01$, $p < .001$, so the program proceeded to the next step.

As shown next to the footnote labeled 'c,' for the third model, beginning salary, previous experience, and months since hire were entered as predictors (months since hire was entered next because it has the second largest semi-partial correlation with current salary). As shown in the row labeled '3,' the change in the R^2 value associated with the addition of months since hire was small but statistically significant, $\Delta R^2 = .01$, $\Delta F(1, 470) = 24.95$, $p < .001$, so the program advanced to the fourth model.

The footnote labeled 'd' under the table indicates that the fourth model included beginning salary, previous experience, months since hire, and educational level as predictors. The change in the R^2 value associated with the addition of educational level was also small but statistically significant, $\Delta R^2 = .007$, $\Delta F(1, 469) = 16.37$, $p < .001$, so the program advanced to the fifth model.

The footnote labeled 'e' under the table indicates that the fifth model included beginning salary, previous experience, months since hire, educational level, and gender as predictors. The change in the R^2 value associated with the addition of gender was once again small but statistically significant, $\Delta R^2 = .003$, $\Delta F(1, 468) = 7.95$, $p = .005$. The model identified as the most superior is always reported at the bottom of the table. Since this is the last model included in the table, we can infer that it is the best model and that the addition of the remaining predictor (minority classification) does not increase R^2 by a significant amount.

Model Statistics

The 'ANOVA' table shown below provides the degrees of freedom, F statistics, and p values for all five of the regression models. Since stepwise regression is used to find the model that best fits the data, we are only interested in the statistics provided in the lowest portion of the table (those associated with Model 5, which SPSS has identified as the best model). As you can see, the regression model that includes beginning salary, previous experience, months since hire, educational level, and gender as predictors is statistically significant, $F(5, 468) = 408.69$, $p < .001$.

ANOVA[a]

Model		Sum of Squares	df	Mean Square	F	Sig.
1	Regression	1.068E+11	1	1.068E+11	1622.118	.000[b]
	Residual	3.109E+10	472	65858997.2		
	Total	1.379E+11	473			
2	Regression	1.094E+11	2	5.472E+10	904.752	.000[c]
	Residual	2.848E+10	471	60476323.3		
	Total	1.379E+11	473			
3	Regression	1.109E+11	3	3.696E+10	642.151	.000[d]
	Residual	2.705E+10	470	57550239.5		
	Total	1.379E+11	473			
4	Regression	1.118E+11	4	2.794E+10	501.450	.000[e]
	Residual	2.614E+10	469	55728306.8		
	Total	1.379E+11	473			
5	Regression	1.122E+11	5	2.244E+10	408.692	.000[f]
	Residual	2.570E+10	468	54914875.0		
	Total	1.379E+11	473			

a. Dependent Variable: Current Salary

b. Predictors: (Constant), Beginning Salary

c. Predictors: (Constant), Beginning Salary, Previous Experience (months)

d. Predictors: (Constant), Beginning Salary, Previous Experience (months), Months since Hire

e. Predictors: (Constant), Beginning Salary, Previous Experience (months), Months since Hire, Educational Level (years)

f. Predictors: (Constant), Beginning Salary, Previous Experience (months), Months since Hire, Educational Level (years), GenderCode

Predictor Statistics

Finally, we will consider the 'Coefficients' table shown on the following page as it contains our regression coefficients and information on each of the predictors.

Coefficients[a]

Model		Unstandardized Coefficients		Standardized Coefficients	t	Sig.
		B	Std. Error	Beta		
1	(Constant)	1928.206	888.680		2.170	.031
	Beginning Salary	1.909	.047	.880	40.276	.000
2	(Constant)	3850.718	900.633		4.276	.000
	Beginning Salary	1.923	.045	.886	42.283	.000
	Previous Experience (months)	-22.445	3.422	-.137	-6.558	.000
3	(Constant)	-10266.629	2959.838		-3.469	.001
	Beginning Salary	1.927	.044	.888	43.435	.000
	Previous Experience (months)	-22.509	3.339	-.138	-6.742	.000
	Months since Hire	173.203	34.677	.102	4.995	.000
4	(Constant)	-16149.671	3255.470		-4.961	.000
	Beginning Salary	1.768	.059	.815	30.111	.000
	Previous Experience (months)	-17.303	3.528	-.106	-4.904	.000
	Months since Hire	161.486	34.246	.095	4.715	.000
	Educational Level (years)	669.914	165.596	.113	4.045	.000
5	(Constant)	-10317.115	3837.191		-2.689	.007
	Beginning Salary	1.723	.061	.794	28.472	.000
	Previous Experience (months)	-19.436	3.583	-.119	-5.424	.000
	Months since Hire	154.536	34.085	.091	4.534	.000
	Educational Level (years)	593.031	166.630	.100	3.559	.000
	GenderCode	-2232.917	792.078	-.065	-2.819	.005

a. Dependent Variable: Current Salary

Once again, we are primarily interested in the statistics provided in the lowest portion of this table, since they pertain to the model that SPSS identified as being the most superior. The program will cease once the addition of a predictor does not significantly increase R^2, and the 'Model Summary,' 'ANOVA,' and 'Coefficients' tables will only show those models that contain significant predictors. Any model that includes a non-significant predictor (e.g., models that include minority classification) will not be shown in these tables. As such, we already know that all of the predictors contained in the final model will be statistically significant.

Nevertheless, we can use the information in the lowest portion of this table to report the relevant statistics for each of our predictors and to construct the equation for the least-squares regression line. Note, however, that the value of the degrees of freedom that we will need to report for each predictor must be found in the 'Model Summary' table shown on page 93 in the last row of the column labeled 'df2'.[21] Let's start by reporting the results of each of the predictors. The results show that beginning salary, $\beta = .79$, $t(468) = 28.47$, $p < .001$, previous experience, $\beta = -.12$, $t(468) = -5.42$, $p < .001$, months since hire, $\beta = .09$, $t(468) = 4.53$, $p < .001$, educational level, $\beta = .10$, $t(468) = 3.56$, $p < .001$, and gender, $\beta = -.07$, $t(468) = -2.82$, $p = .005$, are all significant predictors of current salary.

---

[21]The same value can be found in the bottom portion of the 'ANOVA' table in the row labeled 'Residual.'

Excluded Variables

The final table, displayed below, shows the excluded variables, the variables that were excluded from each of the five regression models.

Excluded Variables[a]

Model		Beta In	t	Sig.	Partial Correlation	Collinearity Statistics Tolerance
1	Educational Level (years)	.172[b]	6.356	.000	.281	.599
	Months since Hire	.102[b]	4.750	.000	.214	1.000
	Previous Experience (months)	-.137[b]	-6.558	.000	-.289	.998
	Minority Classification	-.040[b]	-1.794	.073	-.082	.975
	GenderCode	-.061[b]	-2.482	.013	-.114	.791
2	Educational Level (years)	.124[c]	4.363	.000	.197	.520
	Months since Hire	.102[c]	4.995	.000	.225	1.000
	Minority Classification	-.019[c]	-.869	.385	-.040	.952
	GenderCode	-.088[c]	-3.740	.000	-.170	.771
3	Educational Level (years)	.113[d]	4.045	.000	.184	.516
	Minority Classification	-.024[d]	-1.127	.260	-.052	.950
	GenderCode	-.079[d]	-3.406	.001	-.155	.765
4	Minority Classification	-.024[e]	-1.185	.237	-.055	.950
	GenderCode	-.065[e]	-2.819	.005	-.129	.744
5	Minority Classification	-.033[f]	-1.618	.106	-.075	.930

a. Dependent Variable: Current Salary

b. Predictors in the Model: (Constant), Beginning Salary

c. Predictors in the Model: (Constant), Beginning Salary, Previous Experience (months)

d. Predictors in the Model: (Constant), Beginning Salary, Previous Experience (months), Months since Hire

e. Predictors in the Model: (Constant), Beginning Salary, Previous Experience (months), Months since Hire, Educational Level (years)

f. Predictors in the Model: (Constant), Beginning Salary, Previous Experience (months), Months since Hire, Educational Level (years), GenderCode

This table simply shows the test statistics for each of the predictor variables that were excluded from each of the regression models. The top portion of the table shows that all of the predictor variables, except for beginning salary, were excluded from the model. It also displays what the beta weights, t statistics, and p values for each of these predictors would have been if they had been included in the model.

Once again, the bottom portion of the table is of primary interest because it shows that if minority classification (the only variable that was excluded from the final model) had been included, it would not have been a significant predictor, $\beta = -.03$, $t(467) = -1.62$, $p = .11$, and accordingly would not have increased R^2 by a significant amount. This confirms that the model containing only the five remaining predictors is the most superior model.

Finally, you should be aware that SPSS will also remove predictor variables that no longer significantly improve the model once other predictor variables have been added. For example, if months since hire was no longer a significant predictor after educational level was added, and educational level proved to be a superior predictor in the model relative to months since hire, then the program would have excluded months since hire from subsequent analyses.

Constructing the Equation for the Least-Squares Regression Line

The values listed in the 'Coefficients' table shown below can also be used to construct the least-squares regression line. The equation for a model with five predictor variables is: $Y' = b_1X_1 + b_2X_2 + b_3X_3 + b_4X_4 + b_5X_5 + a$. Alternatively, this formula may be expressed as: $\hat{y} = a + bX_1 + bX_2 + bX_3 + bX_4 + bX_5$.

Coefficients[a]

Model		Unstandardized Coefficients		Standardized Coefficients	t	Sig.
		B	Std. Error	Beta		
1	(Constant)	1928.206	888.680		2.170	.031
	Beginning Salary	1.909	.047	.880	40.276	.000
2	(Constant)	3850.718	900.633		4.276	.000
	Beginning Salary	1.923	.045	.886	42.283	.000
	Previous Experience (months)	-22.445	3.422	-.137	-6.558	.000
3	(Constant)	-10266.629	2959.838		-3.469	.001
	Beginning Salary	1.927	.044	.888	43.435	.000
	Previous Experience (months)	-22.509	3.339	-.138	-6.742	.000
	Months since Hire	173.203	34.677	.102	4.995	.000
4	(Constant)	-16149.671	3255.470		-4.961	.000
	Beginning Salary	1.768	.059	.815	30.111	.000
	Previous Experience (months)	-17.303	3.528	-.106	-4.904	.000
	Months since Hire	161.486	34.246	.095	4.715	.000
	Educational Level (years)	669.914	165.596	.113	4.045	.000
5	(Constant)	-10317.115	3837.191		-2.689	.007
	Beginning Salary	1.723	.061	.794	28.472	.000
	Previous Experience (months)	-19.436	3.583	-.119	-5.424	.000
	Months since Hire	154.536	34.085	.091	4.534	.000
	Educational Level (years)	593.031	166.630	.100	3.559	.000
	GenderCode	-2232.917	792.078	-.065	-2.819	.005

a. Dependent Variable: Current Salary

Once again, we are only interested in the lowest portion of the table, pertaining to model 5. As reviewed in Chapter 4, all of the regression coefficients that we need to construct the equation for the least-squares regression line can be found in the column labeled 'B.' By referring to the rows labeled with the variable names, you should be able to determine that the value of b_1 (the slope for beginning salary) is 1.72, the value of b_2 (the slope for previous experience) is -19.44, the value of b_3 (the slope for months since hire) is 154.54, the value of b_4 (the slope for educational level) is 593.03, and the slope of b_5 (the slope for gender) is -2232.92. Finally, by referring to the row labeled '(Constant),' you should be able to see that the value of a (the intercept) is -10317.12. Now, all we need to do is substitute these values into the equation: $Y' = b_1X_1 + b_2X_2 + b_3X_3 + b_4X_4 + b_5X_5 + a$

$$Y' = 1.72X_1 - 19.44X_2 + 154.54X_3 + 593.03X_4 - 2232.92X_5 - 10317.12$$

The negative values of b_2 and b_5 indicate that the slopes for these variables are inverse. In other words, previous experience and gender show a negative relationship with current salary. This means that individuals with fewer months of previous experience tend to receive higher current salaries[22]. The inverse slope associated with the coded gender variable indicates that women tend to receive lower current salaries.

Using the Regression Equation to Make Predictions

We can now use the regression equation to predict people's current salaries based on their beginning salaries, months of previous experience, months since hire, years of education, and gender. Let's say that Evan was paid a beginning salary of $25,000. He has 24 months of previous experience, was hired 12 months ago, has 18 years of education, and is male. What would we predict his current salary to be? To answer this question, we simply need to substitute these values in for the corresponding Xs in our regression equation and then solve the equation. Since beginning salary is labeled X_1, previous experience X_2, months since hire X_3, educational level X_4, and gender X_5, we need to substitute 25,000 for X_1, 24 for X_2, 12 for X_3, 18 for X_4, and 1 (the numeric code we assigned for males) for X_5, and then solve. Let's go ahead and do that:

$$Y' = 1.7228X_1 - 19.4361X_2 + 154.5362X_3 + 593.0313X_4 - 2232.9171X_5 - 10317.1153$$
$$Y' = 1.7228(25000) - 19.4361(24) + 154.5362(12) + 593.0313(18) - 2232.9171(1) - 10317.1153$$
$$Y' = 42582.50$$

Thus, we would predict Evan's current salary to be $42,582.50.

Reporting the Results

We could now report the following results:

> A stepwise regression analysis was conducted using beginning salary, educational level, months since hire, previous experience, gender, and minority classification to predict current salary. The results revealed a significant five-predictor model, $R^2 = .81$, $F(5, 468) = 408.69$, $p < .001$. Specifically, beginning salary, $\beta = .79$, $t(468) = 28.47$, $p < .001$, previous experience, $\beta = -.12$, $t(468) = -5.42$, $p < .001$, months since hire, $\beta = .09$, $t(468) = 4.53$, $p < .001$, educational level, $\beta = .10$, $t(468) = 3.56$, $p < .001$, and gender, $\beta = -.07$, $t(468) = -2.82$, $p = .005$, were all significant predictors of current salary. Minority classification was not included in the model as its inclusion would not have significantly improved upon its predictive utility, $\beta = -.03$, $t(467) = -1.62$, $p = .11$.

---

[22] This seems odd and may be an artifact that is unique to this sample.

Sign Test

Learning Objectives

In this chapter, you will learn how to analyze data using the sign test. You will learn how to conduct the analysis and interpret and report the results for both two-tailed and one-tailed tests.

The sign test can be used to analyze data from a repeated measures (within-subjects) design with two conditions. It allows you to assess whether there is a significant difference in the direction of responses across the two conditions of an experiment. Specifically, the difference in each participant's scores on the dependent variable across the two conditions is expressed as a plus or a minus sign, and the total number of plus and minus signs are compared to determine whether there is a significant difference. Unlike the t test (described in Chapter 7), it does not carry the assumptions that the data are normally distributed or that the dependent variable was measured on an interval or ratio scale. Since it does not carry these assumptions, it is referred to as a nonparametric test.

NON-DIRECTIONAL HYPOTHESES AND TWO-TAILED SIGN TESTS

Imagine you have just been hired by campus food services to conduct a Pepsi Challenge in order to determine whether students on your campus prefer the taste of Pepsi or Coke. You predict that one brand will be preferred over the other. Since you do not predict which brand of soft drink will be preferred, your hypothesis is considered non-directional and you will need to perform a two-tailed sign test. Prior to conducting the experiment you decide to stick with convention and set alpha at .05.

You go to the Student Union Building and ask 12 students to do a blind taste test. You ask each student to take a sip of Pepsi and a sip of Coke and then to indicate which they prefer. To guard against order effects you use complete counterbalancing. You randomly select six students to take a sip of Pepsi first and a sip of Coke second and six students to take a sip of Coke first and a sip of Pepsi second.

Assume you obtain the results shown below. A '1' indicates that the participant preferred the soft drink and a '2' means they didn't prefer the soft drink. You should be able to see that the participant with the ID code 1 preferred Pepsi while the participant with ID code 4 preferred Coke.

ID	Coke	Pepsi	ID	Coke	Pepsi
1	2	1	7	2	1
2	2	1	8	1	2
3	2	1	9	2	1
4	1	2	10	2	1
5	2	1	11	2	1
6	2	1	12	2	1

Begin by **opening a blank SPSS spreadsheet** and **entering the data** into three columns and 12 rows in the **Data View window** (refer to the Entering Data section on page 7 if you forget how to do this). Note that the top row contains the **variable names**, which must be entered in the **Variable View** window. While you are in Variable View, you should label the values of the Coke and Pepsi variables using the **Values** column. Label **1** as **Preferred** and **2** as **Not Preferred**, and define their scale of **Measure** as **Nominal** (refer to the Creating Variables section on pages 5–7 if you forget how to do this). Once the data are entered, your Data View window should look like the one shown on the following page.

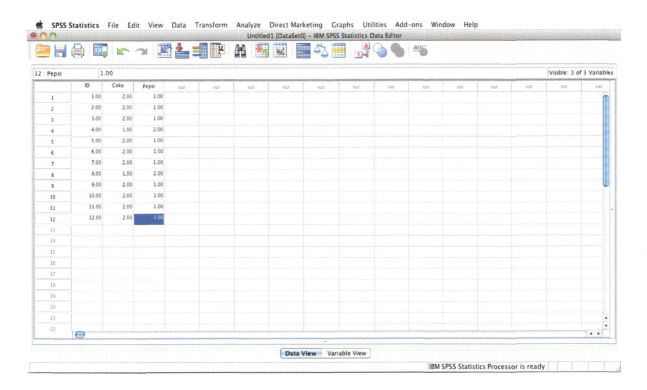

Conducting a Sign Test
(Analyze→Nonparametric Tests→Legacy Dialogs→2 Related Samples)

To initiate the sign test analysis, use the upper toolbar to go to **Analyze→Nonparametric Tests→Legacy Dialogs→2 Related Samples**.

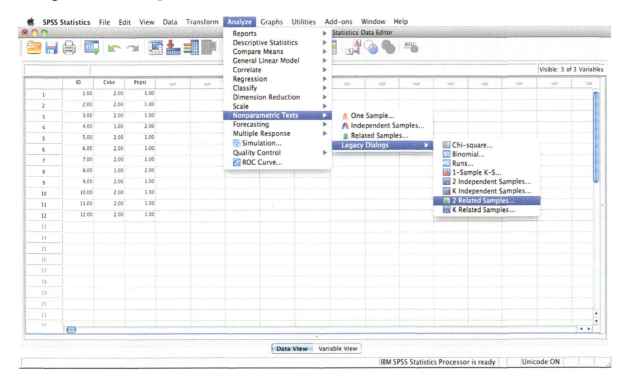

A dialog window labeled 'Two-Related-Samples Tests' will now appear with the variables listed on the left side. Highlight the relevant variables—**Pepsi** and **Coke**—by clicking on each variable once, and move the pair over to the **Test Pairs box** by clicking on the **blue arrow**. **Check** the box labeled **Sign** and **uncheck** the box labeled **Wilcoxon**. Click **OK** to close the dialog window and execute the analysis.

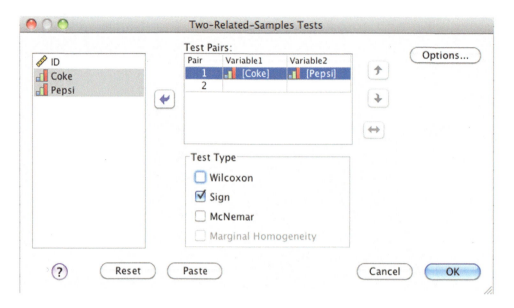

Interpreting the Results

An output window will now appear displaying two tables. We will begin by examining the 'Frequencies' table shown below.

Frequencies

		N
Pepsi – Coke	Negative Differences[a]	10
	Positive Differences[b]	2
	Ties[c]	0
	Total	12

a. Pepsi < Coke

b. Pepsi > Coke

c. Pepsi = Coke

The 'Frequencies' table displays the total number of negative differences (minus signs), the total number of positive differences (plus signs), and the total number of ties. Since the signs are meaningless unless you know which variable was subtracted from which, the superscripts 'a,' 'b,' and 'c' presented in the table are required to clarify what a negative difference represents and what a positive difference represents. The footnotes below the table clearly show that a negative difference (superscript a) indicates that Pepsi has a lower value than Coke, and that a positive difference (superscript b) indicates that Coke has a lower value than Pepsi. We used lower numbers (1s) to indicate a preference for the brand of pop, therefore a negative difference indicates a preference for Pepsi and a positive difference indicates a preference for Coke. Thus, 10 people indicated a preference for Pepsi and only two indicated a preference for Coke.

The output window also contains the 'Test Statistics' table shown below.

Test Statistics[a]

	Pepsi – Coke
Exact Sig. (2-tailed)	.039[b]

a. Sign Test

b. Binomial distribution used.

The 'Test Statistics' table displays the obtained p value for a two-tailed sign test in the row labeled 'Exact Sig. (2-tailed).' Since our hypothesis was non-directional (we did not predict which brand would be preferred), this value is appropriate. The p value represents the probability we would obtain these results or results more extreme based on chance alone (the probability we would obtain 10 or more minuses from 12 participants if the two brands of soft drinks are actually equally preferred in the population). The table shows that the p value is .04. Thus, the probability that this particular result (10 negative differences) or results more extreme would be obtained based on chance alone is .04.

Reporting the Results

Since $p < .05$, we conclude that the two brands of soft drinks are not equally preferred (i.e., that one brand is preferred over the other). Moreover, since the number of negative differences outweighs the number of positive differences, and negative differences represent a preference for Pepsi, we can further conclude that Pepsi is preferred over Coke, in taste tests.[23] We could report these results simply by stating the following:

> A sign test revealed a significant difference in soft drink preference, with 10 of the 12 participants preferring Pepsi to Coke ($p = .04$).

DIRECTIONAL HYPOTHESES AND ONE-TAILED SIGN TESTS

In the previous example, we used a non-directional hypothesis and two-tailed test. For this next example, we will use a directional hypothesis (we will predict the direction of effect) and a one-tailed test.

Imagine you have been hired by a drug company to test the effectiveness of a new sleep medication called Zzz-Aid. Specifically, the company wants to know whether Zzz-Aid is *more* effective (i.e., puts people to sleep faster) than a placebo. Prior to conducting the study, you decide to set alpha at .05 (one-tailed).

---

[23] While the data contained in this guide are entirely fictional, some interesting research on the Pepsi challenge was conducted because Pepsi typically comes out on top in taste tests but Coke tends to sell more product than Pepsi. Pepsi is sweeter than Coke and the results of research indicate that people prefer sweeter drinks (Pepsi) when taking only a sip, but when drinking an entire can they prefer the less sweet Coke. Unfortunately, this discovery was made after Coke spent millions on the sweeter tasting "New Coke" in the 80s—a huge failure for them! For those of you who have taken a research methods course, you may recognize this as an issue of external validity. Taste tests have problems with external validity because in real life people tend to drink more than a sip.

You randomly sample 18 individuals from the community and invite each to your sleep lab for two nights. You give each individual Zzz-Aid on one night and a placebo on the other night. To guard against order effects, you use complete counterbalancing. You randomly select nine participants to take Zzz-Aid on the first night and a placebo on the second night, and nine participants to take the placebo on the first night and Zzz-Aid on the second night. Participants are not told which night they are given the placebo or the drug. You measure and record the number of minutes it takes each participant to fall asleep on each night. Assume you obtained the data shown below.

ID	ZzzAid	Placebo	ID	ZzzAid	Placebo
1	12	21	10	16	19
2	9	16	11	10	15
3	11	8	12	15	9
4	21	36	13	10	22
5	17	28	14	28	32
6	22	20	15	8	12
7	18	29	16	20	18
8	11	22	17	12	12
9	10	21	18	8	8

Begin by **opening a blank SPSS spreadsheet** and **entering the data** into three columns and 18 rows in the Data View window. Once again, note that the top row contains the **variable names**, which must be entered in the **Variable View** window. While you are in the Variable View window, you should define the scale of measure of **ID** as **Nominal**, and the scales of measure of ZzzAid and Placebo as **Scale**. Once the data are entered, your Data View window should look like the one displayed below.

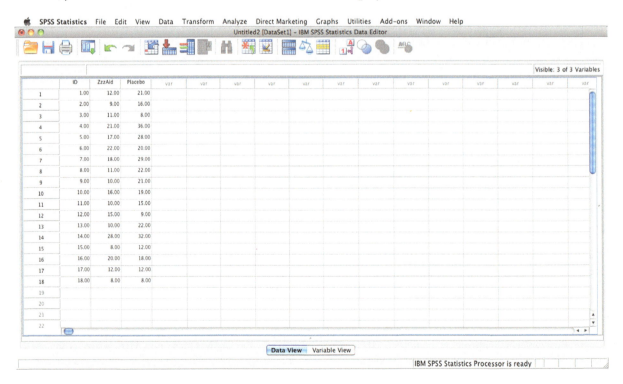

Conducting a Sign Test
(Analyze→Nonparametric Tests→Legacy Dialogs→2 Related Samples)

The method for conducting a one-tailed sign test is the same as that for conducting a two-tailed sign test. Using the upper toolbar, go to **Analyze→Nonparametric Tests→Legacy Dialogs→2 Related Samples**. Using the 'Two-Related-Samples Test' dialog window shown below, highlight the relevant variables— **ZzzAid** and **Placebo**—by clicking on each variable once. Move the pair over to the **Test Pairs box** by clicking on the **blue arrow**. **Check** the box labeled **Sign** and **uncheck** the box labeled **Wilcoxon**. Click **OK**.

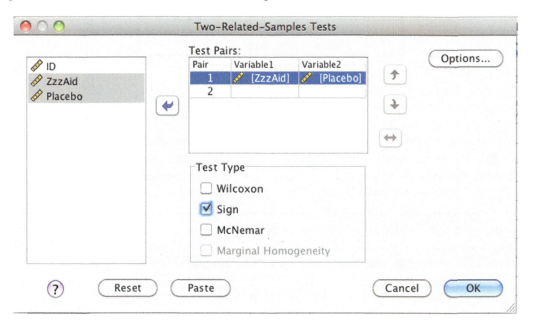

Interpreting the Results

We will once again start by examining the 'Frequencies' table that appears in the output window.

Frequencies

		N
Placebo – ZzzAid	Negative Differences[a]	4
	Positive Differences[b]	12
	Ties[c]	2
	Total	18

a. Placebo < ZzzAid

b. Placebo > ZzzAid

c. Placebo = ZzzAid

From this table you should be able to determine that there are four negative differences, 12 positive differences, and two ties. The footnotes below the table show that a negative difference represents lower scores in the placebo condition than in the Zzz-Aid condition. Since the data represent the number of minutes it took for participants to fall asleep, this means that four participants fell asleep faster when they took the placebo. The footnotes below the table also show that a positive difference represents lower scores in the Zzz-Aid condition. Therefore, we can determine that 12 participants fell asleep faster when they took Zzz-Aid. The two ties mean that for two participants there was no difference in the number of minutes it took them to fall asleep. The sign test will not tolerate ties, so it simply discards the data from ties. Thus, our sample size is now reduced to 16 participants ($N = 16$).

Test Statistics[a]

	Placebo – ZzzAid
Exact Sig. (2-tailed)	.077[b]

a. Sign Test

b. Binomial distribution used.

The 'Test Statistics' table shown above displays the p value for a two-tailed test in the row labeled 'Exact Sig. (2-tailed)'. You should recall from Chapter 3 that two-tailed tests are used when the hypothesis is non-directional. However, for this example, we have a directional hypothesis and therefore we should use a one-tailed test. SPSS does not provide an option to run a one-tailed sign test, but it is a simple process to manually convert the p value. To obtain the p value for a one-tailed test, simply divide the p value for the two-tailed test in half. For added precision, you should use a p value that has been rounded to at least four decimal places. By clicking repeatedly on the value in the table you should find that the p value rounded to four decimal places is .0768. Therefore, our p value for a one-tailed test is:

$$p = .0768/2 = .0384$$
$$p = .04$$

Remember that when we use a directional hypothesis we can only reject the null hypothesis if the results turn out in the predicted direction. We predicted that Zzz-Aid would decrease the number of minutes it takes participants to fall asleep. As described above, this is represented as a positive difference. Since the number of positive differences outweighed the number of negative differences, we know the results worked out in the predicted direction. Moreover, since $p < .05$ (one-tailed), we can conclude that the difference is statistically significant.

Reporting the Results

We could report these results simply by stating:

A one-tailed sign test revealed that significantly more participants fell asleep faster after taking Zzz-Aid relative to a placebo ($p = .04$).

Predicting the Wrong Direction

You should note that if the results had worked out in the opposite direction to what we predicted and there were instead 12 negative differences and four positive differences (meaning that for most people Zzz-Aid increased the number of minutes it takes to fall asleep), then the displayed p value would be the same but we would have to make a further adjustment to it by subtracting the halved p value from 1.

$$p = 1 - (.0768/2)$$
$$p = 1 - .0384$$
$$p = .96$$

Therefore, in this case, we would report the p value as .96 and since this value is higher than alpha (.05), we would have to conclude that the results are not statistically significant. This once again demonstrates the risk of using a directional hypothesis and the importance of examining the direction of effect before reaching a conclusion about the significance of the results of a one-tailed test.

t Tests

Learning Objectives

In this chapter, you will learn how to analyze data using single-sample *t* tests, paired-samples *t* tests, and independent-samples *t* tests. For each, you will learn how to conduct the analysis and interpret and report the results for both non-directional and directional hypotheses. You will also learn to compute the confidence intervals and the effect size indicator Cohen's *d* for all three types of tests.

SINGLE-SAMPLE *t* TEST

The single-sample *t* test, alternatively referred to as the one-sample *t* test, is used to determine whether the mean of a sample (*M*) is significantly different from the mean of a population (μ). As such, it can be used to infer whether a sample belongs to a population or is unique from a population.

According to Statistics Canada's census results, the mean weight of Canadian women was 153 lbs in 2005. Imagine you are a health researcher interested in examining whether there has been a significant change in the mean weight of Canadian women since 2005. While obesity rates are currently at an all time high, you also recognize that many Canadians are beginning to adopt healthier lifestyles. As such, you decide to use a non-directional hypothesis and simply predict that there has been a significant change in the mean weight of Canadian women since 2005. Prior to collecting any data, you decide to stick with the convention and set alpha at .05.

Assume you randomly sample 30 women from the general population of Canada and obtain the following data on their weights (in lbs):

ID	Weight	ID	Weight	ID	Weight
1	165	11	149	21	129
2	110	12	136	22	117
3	107	13	212	23	152
4	163	14	157	24	183
5	195	15	132	25	169
6	159	16	128	26	145
7	168	17	178	27	265
8	145	18	210	28	162
9	125	19	159	29	179
10	198	20	195	30	285

Begin by **opening a blank SPSS spreadsheet** and **entering** the 30 women's **ID codes** and **weights** into two columns in the **Data View window** (refer to the Entering Data section on page 7 if you forget how to do this). Note that the top row contains the **variable names**, which must be entered in the **Variable View** window. While you are in the Variable View window, you should define the properties of the variables. **Label ID** as **Participant ID Codes** and **Weight** as **Weight in Pounds** and set the scale of **Measure** to **Nominal** for **ID**, and to **Scale** for **Weight** (refer to the Creating Variables section on pages 5–7 if you forget how to do this). Once the data are entered your Data View window should look like the one shown on the following page.

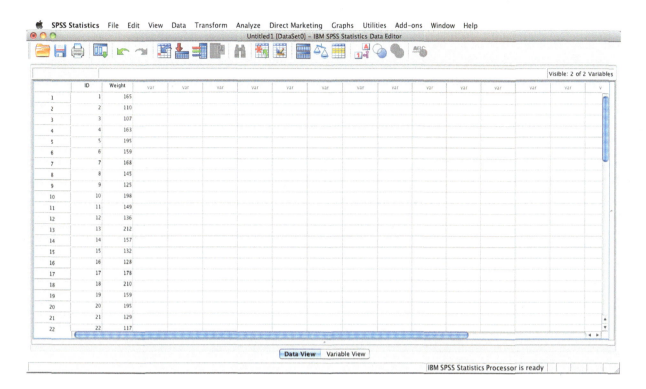

Conducting a Single-Sample *t* Test
(Analyze→Compare Means→One-Sample T Test)

Since we want to compare a sample mean to a population mean to determine whether there is a significant difference, we will need to conduct a single-sample *t* test. To conduct the analysis, use the upper toolbar to go to **Analyze→Compare Means→One-Sample T Test**.

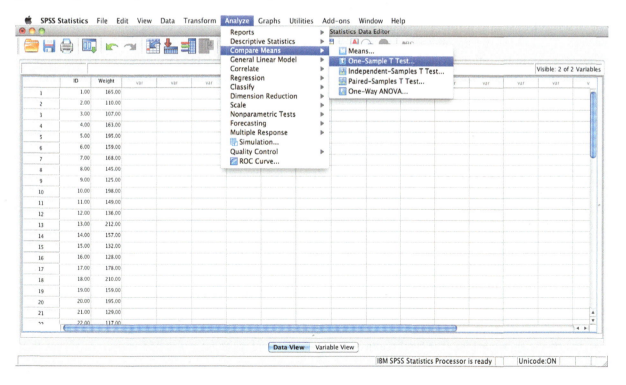

A 'One-Sample T Test' dialog window like the one shown below will now appear. Highlight the relevant variable—**Weight**—by clicking on the variable label, and move it over to the **Test Variable(s) box** by clicking the **blue arrow**. Next, type the population mean—**153**—into the **Test Value box**. Click **OK** to close the dialog window and execute the analysis.

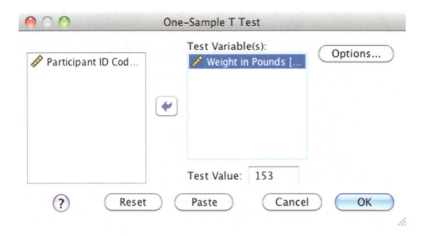

Interpreting the Results

An output window will now appear displaying two tables. We will begin by examining the 'One-Sample Statistics' table shown below.

One-Sample Statistics

	N	Mean	Std. Deviation	Std. Error Mean
Weight in Pounds	30	165.9000	40.88449	7.46445

The first column simply displays the label for the variable you are considering in the analysis. The column labeled 'N' displays the number of participants in the sample, the column labeled 'Mean' displays the mean of the sample, and the column labeled 'Std. Deviation' displays the standard deviation of the sample. Finally, the column labeled 'Std. Error Mean' displays the standard deviation of the sampling distribution of the mean, which is also known as the standard error of the mean (abbreviated *SEM*).

Next, we will consider the 'One-Sample Test' table shown below.

One-Sample Test

	Test Value = 153					
			Sig. (2-tailed)	Mean Difference	95% Confidence Interval of the Difference	
	t	df			Lower	Upper
Weight in Pounds	1.728	29	.095	12.90000	-2.3665	28.1665

The primary results of the analysis are presented in this table. The mean of the population is provided at the very top of the table next to the label 'Test Value = .' You can use this to confirm that you correctly entered the mean of the population. The column labeled 't' provides the obtained *t* value. From this we can determine that *t* = 1.73. The next column, labeled 'df,' displays the degrees of freedom; it shows that we have 29 degrees of freedom. The column labeled 'Sig. (2-tailed)' shows the precise *p* value. Since this *p* value is greater than the alpha level of .05 that we set prior to collecting the data, we will need to conclude that the difference between the sample mean and population mean is not statistically significant. The column labeled 'Mean Difference' displays the difference between the sample mean and the population mean. This value indicates that our sample has a mean weight that is 12.90 lbs higher than the mean weight of the population.

The last two columns of the 'One-Sample Test' table display the lower and upper limits of the 95% confidence interval for the *mean difference*. The 95% confidence interval for the mean difference provides a range of values that you can be 95% certain contains the true difference between the sample and population means. Thus, the 95% confidence interval for the mean difference displayed above indicates that we can be 95% certain that the interval −2.37 lbs to 28.17 lbs contains the true difference in Canadian women's mean weight from 2005 to now. Since that interval contains 0 (it contains the possibility that there is no difference in the mean weights), we must conclude that the difference is not statistically significant.

Calculating Confidence Intervals for the Population Mean

Researchers are often interested in obtaining the confidence interval for the *population mean* (rather than the confidence interval for the mean difference). The 95% confidence interval for the population mean provides a range of values that you can be 95% certain contains the true population mean. This is a subtle but important distinction. Although SPSS only provides the confidence interval for the mean difference, the confidence interval for the population mean can easily be hand calculated. To calculate the 95% confidence interval for the population mean we will need the sample mean (M), the standard error of the mean (SEM), and the critical *t* value (t_{crit}) for an alpha of .05 and 29 degrees of freedom. As described above, the sample mean and the standard error of the mean can both be found in the 'One-Sample Statistics' table. To determine the critical *t* value you will need to refer to a table of Critical Values of Student's *t* distribution.[24] For a two-tailed test with alpha of .05 and 29 degrees of freedom, the critical *t* value is 2.0452. The formulas for calculating the lower and upper limits of the confidence interval for the population mean are provided below. Let's go ahead and practice hand calculating the confidence interval for the population mean.

$$\mu_{\text{lower}} = M - SEM(t_{\text{crit}})$$
$$\mu_{\text{upper}} = M + SEM(t_{\text{crit}})$$

$$\mu_{\text{lower}} = 165.9000 - 7.4645(2.0452)$$
$$\mu_{\text{lower}} = 150.63$$

$$\mu_{\text{upper}} = 165.9000 + 7.4645(2.0452)$$
$$\mu_{\text{upper}} = 181.17$$

Since the mean weight of the population of Canadian women in 2005 (153 lbs) is contained within this interval, we will once again need to conclude that there has not been a significant change in the mean weight of Canadian women since 2005. Note that using either the traditional (*p* value) method or the alternative confidence interval method we will always arrive at the same conclusion about the significance of our findings.

---

[24] A table of critical values for the *t* distribution can be found at: easycalculation.com/statistics/t-distribution-critical-value-table.php

Calculating the Effect Size (Cohen's d)

Although SPSS does not contain an option to compute an effect size indictor, APA guidelines urge us to report such indicators. Cohen's *d* is traditionally used as an effect size indicator for *t* tests. It is very easy to hand calculate. The formula for Cohen's *d* for the single-sample *t* test is:

$$d = \frac{|M - \mu|}{SD}$$

The formula shows that the Cohen's *d* value is computed by dividing the absolute difference in the sample mean (*M*) and population mean (μ), by the standard deviation of the sample (*SD*). Since all of these values are provided in the SPSS output, we can easily compute the effect size. As shown in the tables described above, the mean of the sample is 165.90, the mean of the population is 153.00, and the standard deviation of the sample is 40.88. Plugging these values (rounded to four decimal places) into the formula we get:

$$d = \frac{|165.9000 - 153.0000|}{40.8845} = 0.32$$

Cohen's *d* values of 0.20 are generally considered small, values of 0.50 are generally interpreted as medium, while values of 0.80 and higher are generally considered large. More specifically, the Cohen's *d* value for a single-sample *t* test is interpreted as the number of standard deviation units the sample mean is from the population mean. Therefore, our Cohen's *d* value of 0.32 indicates that the effect is small. More specifically, this value indicates that our sample mean is 0.32 standard deviation units from our population mean.

Reporting the Results

The results of all *t* tests are reported using the following format: $t(df)$ = #.##, *p* = .##, *d* = #.##, ##% CI [#.##, #.##]. Note that since Cohen's *d* values can be greater than 1, a leading zero should be used before the decimal place when reporting values below 1 and the *d* should be italicized. It is also customary to report the population mean as well as the mean and the standard deviation of the sample. Thus, for the analysis we conducted we would report the following:

> A single-sample *t* test revealed that the mean weight of today's sample of Canadian women (*M* = 165.90, *SD* = 40.88) is not significantly different than the mean weight of the population of Canadian women in 2005 (μ = 153), $t(29)$ = 1.73, *p* = .09, *d* = 0.32, 95% CI [150.63, 181.17].

One-Tailed Single-Sample *t* Test

Let's now assume that you are a dietician who has growing concerns about the impact of the rapid growth of the fast food industry on Canadian women's weight. As such, you are only interested in whether the mean weight of Canadian women has *increased* since 2005. Since your hypothesis is now directional, a one-tailed test is appropriate. Assume that prior to collecting any data you decide to set alpha at .05 (one-tailed). You randomly sample the same 30 Canadian women whose data appear at the beginning of this chapter.[25]

SPSS does not provide an option to conduct one-tailed *t* tests, so directional hypotheses do not influence the way we conduct the analysis. The only thing that differs with directional hypotheses is the manner in which we report the *p* value displayed in the 'One-Sample Test' table. The *p* value provided in the table is

---

[25] Note that this directional hypothesis would have to have been made before collecting any data. It is scientific cheating to switch from a non-directional to a directional hypothesis after seeing the results! In the long run, practicing this type of cheating will inflate your Type I error rate to 7.5%.

for a two-tailed test (non-directional hypothesis). To adjust it for our one-tailed test (directional hypothesis), all we need to do is divide it in half. Let's go ahead and do that:

$$p = .0946/2$$
$$p = .047$$

Thus, our *p* value for a one-tailed test is .047.[26] Before we reach any conclusions, we need to examine whether our results are in the predicted direction. Our obtained sample mean (shown in the 'One-Sample Statistics' table) is 165.90, and our population mean (shown at the very top of the 'One-Sample Test' table) is 153. Since our sample mean is higher than our population mean, we know that the results are in the predicted direction.

Reporting the Results

In this scenario of a one-tailed single-sample *t* test, we would report the following results:

> A one-tailed single-sample *t* test revealed that the mean weight of the sample of Canadian women ($M = 165.90$, $SD = 40.88$) is significantly higher than the mean weight of the population of Canadian women in 2005 ($\mu = 153$), $t(29) = 1.73$, $p = .047$ (one-tailed), $d = 0.32$.

Predicting the Wrong Direction

It is important to note what would change if we had made a prediction in the wrong direction. Let's assume we had predicted a *decrease* in the mean weight of Canadian women and we obtained the same results we've been considering here (showing an *increase* in the mean weight of Canadian women). In this case, our calculation of the *p* value would change. Rather than simply dividing the *p* value in half, as we did when the results worked out in the predicted direction, we would now need to subtract the divided *p* value from a value of 1. Let's go ahead and do that:

$$p = 1 - (.0946/2)$$
$$p = 1 - .0473$$
$$p = .95$$

In this case, we would report the following: $t(29) = 1.73$, $p = .95$, $d = .32$. Since $p > .05$ in this scenario, we would need to conclude that there has not been a significant decrease in the mean weight of Canadian women since 2005.

PAIRED-SAMPLES *t* TEST

The paired-samples *t* test (alternatively referred to as the correlated groups *t* test) is used to analyze data from a repeated measures (within-subjects) design with two conditions. Specifically, it is used to determine whether the means of two conditions (containing the same participants) differ significantly.

Assume you are a researcher interested in the acute effects of THC (i.e., the effects of being high on marijuana) on memory test performance. You use a non-directional hypothesis and simply predict that THC will affect memory test performance. Prior to conducting the study you decide to set alpha at .05.

Assume that you randomly select and invite 20 individuals to your lab on two separate days. Each individual is given a joint containing THC (i.e., the active ingredient in marijuana) to smoke on one day and

---

[26] Since the *p* value of .047 rounds to .05, we will report the value rounded to three decimal places. This way our readers/reviewers will know that our value was indeed under .05 (and not .051, which would not be significant).

a placebo joint that doesn't contain any THC to smoke on the other day. To guard against order effects, you use complete counterbalancing. Half of the participants smoke the joint containing THC on the first day and the placebo joint on the second day, and the other half of the participants smoke the placebo joint on the first day and the joint containing THC on the second day. Participants are not informed which day they smoke the joint that contains the THC. After smoking each of the joints, participants are given a memory test. They are read a list of 15 words and are asked to recall as many words from the list as possible. You measure the number of words each participant correctly recalls.

Assume you obtain the data shown below:

ID	THC	Placebo	ID	THC	Placebo
1	7	8	11	6	8
2	5	7	12	5	7
3	4	5	13	9	12
4	12	10	14	12	12
5	8	9	15	15	12
6	9	11	16	7	9
7	11	13	17	5	7
8	6	5	18	9	11
9	7	9	19	10	8
10	8	8	20	5	6

Begin by **entering the data** into a **blank SPSS spreadsheet**. Note that the top row contains the **variable names**, which must be entered in the **Variable View** window. While you are in the Variable View window, you should define the properties of the variables. Set the scale of measure to **Nominal** for **ID**, and to **Scale** for the variables **THC** and **Placebo**. The remaining data must be entered into three columns and 20 rows in the Data View window. Keep in mind that different variables must be entered into separate columns, while the data for each of the participants must be entered in the rows (therefore each row must represent one participant's data on all of the variables). Once the data are entered, your Data View window should look like the one displayed below.

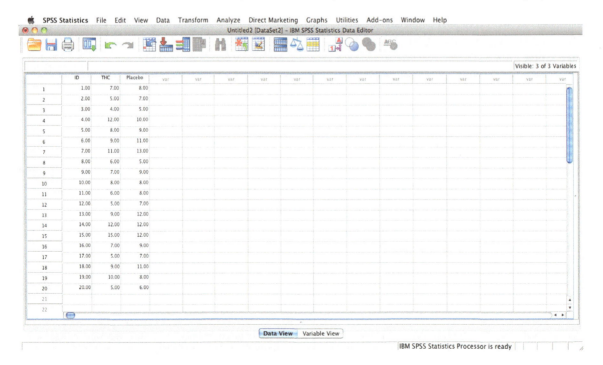

Conducting a Paired-Samples *t* Test
(Analyze→Compare Means→Paired-Samples T Test)

To conduct a paired-samples *t* test, you need to go to **Analyze→Compare Means→Paired-Samples T Test**.

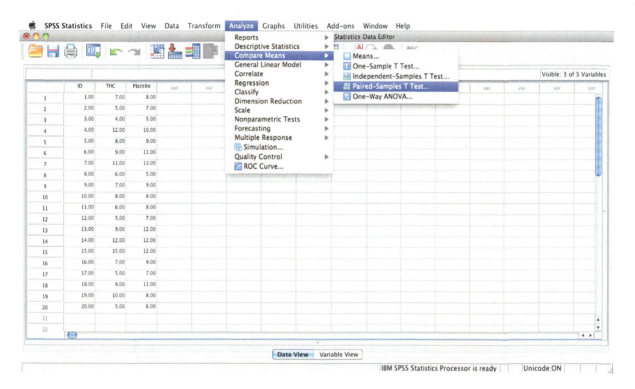

A 'Paired-Samples T Test' dialog window like the one shown below will now appear. Highlight the relevant variables—**THC** and **Placebo**—by clicking on each of the variable names and then click on the **blue arrow** to move them over to the **Paired Variables box**. Click **OK** to initiate the analysis.

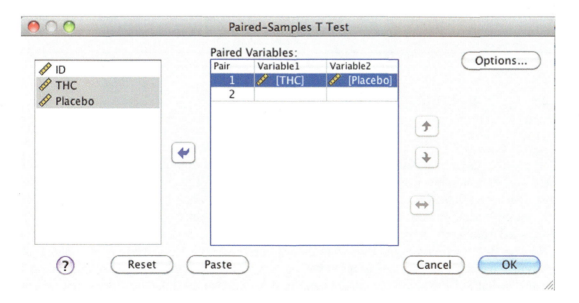

Interpreting the Results

A series of tables will appear in the output window. We will first examine the 'Paired Samples Statistics' table shown below.

Paired Samples Statistics

		Mean	N	Std. Deviation	Std. Error Mean
Pair 1	THC	8.0000	20	2.90191	.64889
	Placebo	8.8500	20	2.39022	.53447

This table provides some basic descriptive statistics. The first column in the table displays the names of the conditions that are being compared. The column labeled 'Mean' presents the mean for each condition. You should be able to see that participants were able to recall a mean of 8.00 words in the THC condition and a mean of 8.85 words in the placebo condition. The column labeled 'N' displays the number of participants in each condition. Since we are considering data from a repeated measures (within-subjects) design, there will always be an equal number of participants in each of the conditions. The column labeled 'Std. Deviation' lists the standard deviation for each condition. Finally, the column labeled 'Std. Error Mean' provides the standard error of the mean for each condition.

The next table we will examine is the 'Paired Samples Correlations' table shown below.

Paired Samples Correlations

		N	Correlation	Sig.
Pair 1	THC & Placebo	20	.819	.000

This table displays the correlation between the scores in the two conditions. You learned all about correlation in Chapter 3, so you should be able to see by looking at this table that there is a large positive correlation, $r(18) = .82$, $p < .001$, between the number of words participants were able to recall after smoking a joint containing THC and the number of words they were able to recall after smoking a placebo joint.

Finally, your output window will display the 'Paired Samples Test' table shown below.

Paired Samples Test

		Paired Differences							
					95% Confidence Interval of the Difference				
		Mean	Std. Deviation	Std. Error Mean	Lower	Upper	t	df	Sig. (2-tailed)
Pair 1	THC – Placebo	-.85000	1.66307	.37187	-1.62834	-.07166	-2.286	19	.034

This table provides us with the results of the *t* test. Once again, the first column displays the conditions being compared and indicates which scores were subtracted from which. The column shows 'THC – Placebo' indicating that scores in the placebo condition were subtracted from scores in the THC condition. The next column, labeled 'Mean,' presents the mean of the difference scores. Since scores in the placebo condition were subtracted from scores in the THC condition, the negative difference score (−0.85) indicates that scores in the THC condition were lower than scores in the placebo condition. More precisely, the value indicates that on average participants were able to recall 0.85 fewer words after smoking the joint containing THC. The column labeled 'Std. Deviation' provides the standard deviation of the difference scores, and the column labeled 'Std. Error Mean' displays the standard error of the mean, which is the estimated standard deviation of the sampling distribution of the mean difference scores.

The section of the table labeled '95% Confidence Interval of the Difference' provides the lower and upper limits of the 95% confidence interval for the difference in means. As shown in the table, this interval is −1.63 to −0.07. This indicates that we can be 95% certain that the interval −1.63 to −0.07 contains the true acute effect of THC on memory test performance. In other words, we can be 95% certain that being high on marijuana decreases memory test performance by 0.07 to 1.63 recalled words. Since the confidence interval does not cross 0, we can conclude that THC has a significant acute effect on memory test performance.

The column labeled 't' displays the obtained *t* value, the column labeled 'df' gives the degrees of freedom, and the column labeled 'Sig. (2-tailed)' provides the *p* value for a two-tailed test. Since *p* < .05, we can once again conclude that THC has a significant acute effect on memory test performance.

Calculating the Effect Size (Cohen's d)

Again, since SPSS does not contain an option to compute the effect size indicator Cohen's *d*, we must calculate it by hand using the information provided in the output. The formula for computing Cohen's *d* for a paired-samples *t* test is:

$$d = \frac{|M_{\mathrm{D}}|}{SD_{\mathrm{D}}}$$

The formula shows that the Cohen's *d* value is computed by dividing the absolute value of the mean of the difference scores (M_{D}) by the standard deviation of the difference scores (SD_{D}). These values are reported in the 'Paired Samples Test' table displayed above. As reviewed previously, the mean of the difference scores is −0.85 (so the absolute value is 0.85), and the standard deviation of the difference scores is 1.66. Plugging these values (rounded to four decimal places) into the formula we get:

$$d = \frac{|-0.8500|}{1.6631} = 0.51$$

The Cohen's *d* value of 0.51 indicates that there is a medium-sized effect of THC on memory test performance. More specifically, it indicates that the mean of the THC condition is 0.51 standard deviation units lower than the mean of the placebo condition.

Reporting the Results

Bringing it all together, we can now report:

> A paired-samples *t* test revealed that the mean number of words participants could recall after smoking the joint containing THC (*M* = 8.00, *SD* = 2.90) was significantly lower than the mean number of words they could recall after smoking the placebo joint (*M* = 8.85, *SD* = 2.39), *t*(19) = −2.29, *p* = .03, *d* = 0.51, 95% CI [−1.63, −0.07].

Calculating the 99% Confidence Intervals for the Difference in Means

By default, SPSS provides the 95% confidence interval because it is the most commonly used confidence interval. However, you can change this setting and request any confidence interval you'd like. We'll practice by changing the setting to give us the second most frequently used confidence interval, the 99% confidence interval (this interval corresponds to an alpha of .01).

To change the confidence interval, go to **Analyze→Compare Means→Paired-Samples T Test**. The variables—**THC** and **Placebo**—should still appear in the **Paired Variables box** (if they do not, then place them in that box using the blue arrow). Next, click the **Options box**. A 'Paired-Samples T Test: Options' window like the one shown below will now appear. Replace the value of 95 with a value of **99** by simply typing in the box. Press **Continue** to close that dialog window and press **OK** to run the analysis.

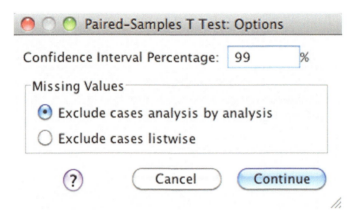

The same three tables will appear in the output window. The only difference between these tables and the ones produced previously is highlighted in the 'Paired Samples Test' table shown below. You should note that nothing else in the tables has changed.

Paired Samples Test

| | | Paired Differences | | | | | | |
| | | | | | 99% Confidence Interval of the Difference | | | | |
	Mean	Std. Deviation	Std. Error Mean	Lower	Upper	t	df	Sig. (2–tailed)
Pair 1 THC – Placebo	-.85000	1.66307	.37187	-1.91390	.21390	-2.286	19	.034

The table now displays the 99% confidence interval for the difference in means. According to this confidence interval, we can be 99% certain that the interval −1.91 to 0.21 contains the true acute effect of THC on memory test performance. Since this confidence interval contains the possibility that THC has no effect on memory test performance (the interval crosses 0), we must conclude that being high on marijuana does not have a significant effect on memory test performance. Note that we would have reached this same conclusion using the traditional p value method and an alpha of .01 (the alpha level that corresponds to the 99% confidence interval) because $p = .03 > .01$.

Reporting the Results

The results using an alpha of .01 and the 99% confidence interval would be reported in the following manner:

> Using an alpha of .01, a paired-samples t test revealed that the mean number of words participants could recall after smoking the joint containing THC ($M = 8.00$, $SD = 2.90$) was not significantly different than the mean number of words they could recall after smoking the placebo joint ($M = 8.85$, $SD = 2.39$), $t(19) = -2.29$, $p = .03$, $d = 0.51$, 99% CI [−1.91, 0.21].

One-Tailed Paired-Samples *t* Test

Now assume you had originally made a directional hypothesis and predicted that THC would *decrease* memory test performance. In this scenario, you would need to conduct a one-tailed paired-samples *t* test. The method of conducting the analysis would remain the same, and the tables in the output would remain the same. The only thing that would change in this scenario is the reported *p* value, which would once again be cut in half. The *p* value for the two-tailed test was .0339, so the *p* value for a one-tailed test would be:

$$p = .0339/2$$
$$p = .02$$

Since the mean of the THC condition was lower than the mean of the placebo condition, we know the results are in the predicted direction.

Reporting the Results

In this scenario of a one-tailed paired-samples *t* test (with alpha of .05), we would report the following:

A one-tailed paired-samples *t* test revealed that the mean number of words participants could recall after smoking the joint containing THC ($M = 8.00$, $SD = 2.90$) was significantly lower than the mean number of words they could recall after smoking the placebo joint ($M = 8.85$, $SD = 2.39$), $t(19) = -2.29$, $p = .02$ (one-tailed), $d = 0.51$.

Predicting the Wrong Direction

Again, if the results had worked out in the opposite direction to which we predicted and showed that memory test scores were actually higher in the THC condition relative to the placebo condition, then we would need to further adjust the *p* value. Rather than simply dividing the *p* value in half as we did when the results worked out in the predicted direction, we would now need to subtract the adjusted *p* value from a value of 1.

$$p = 1 - (.0339/2)$$
$$p = 1 - .0170$$
$$p = .98$$

In this case, we would report the following: $t(19) = -2.29$, $p = .98$, $d = 0.51$. Since $p > .05$ in this scenario, we would need to conclude that THC does not significantly decrease memory test performance.

INDEPENDENT-SAMPLES *t* TEST

The independent-samples *t* test is also used to determine whether the means of two conditions differ significantly. In contrast to the paired-samples *t* test, it is used in conjunction with an independent groups (between-subjects) design in which the two conditions contain different groups of participants.

A previous president of Harvard University once publically speculated that one reason why women are underrepresented in the sciences is because of a "different availability of aptitude at the high end." In other words, he claimed that women are underrepresented in the sciences because they do not have the innate abilities necessary to excel in them. Let's assume you are a researcher who is infuriated by this statement because you think statements like these adversely affect women.[27] As such, you decide to examine the influence of claims like this one on women's actual math test performance. Prior to collecting any data you decide to set alpha at .05.

You invite 20 women to your lab and you randomly assign each woman to one of two separate conditions. Women in both conditions are told to wait in a waiting area where they "accidently overhear" a staged conversation between two professors. Women in condition 1 (the genetics condition) overhear the professors discussing a new paper revealing evidence that a math gene has been discovered that is sex-linked and commonly deficient in women and that this may account for why women are underrepresented in the sciences. Women in condition 2 (the socialization condition) overhear the professors discussing a new paper revealing evidence that women only tend to be underrepresented in the sciences because they are raised in a society where they are regarded as being inferior at math. After participants overhear one of these staged conversations they are asked to complete a math test containing 20 questions. You hypothesize that women's math test performance will be affected by the message they overhear.

Assume you record the following data:

ID	Message	MathScore	ID	Message	MathScore
1	1	15	11	1	8
2	2	18	12	2	20
3	1	12	13	1	12
4	2	13	14	2	16
5	1	11	15	1	14
6	2	10	16	2	11
7	1	15	17	1	9
8	2	17	18	2	16
9	1	14	19	1	16
10	2	19	20	2	20

Begin by **entering the data** into a **blank SPSS spreadsheet**. Note that the top row contains the **variable names**, which must be entered in the **Variable View** window. While you are in Variable View, use the **Values** column to identify message **1** as **Genetics** and message **2** as **Socialization**. Set the scale of **Measure** to **Nominal** for the variables **ID** and **Message**, and to **Scale** for the variable **MathScore**. The remaining data must be entered into three columns and 20 rows in the Data View window. Keep in mind that different variables must be entered into separate columns, while the data for each of the participants must be entered in the rows (therefore each row must represent one participant's data on all of the variables). Once the data are entered, your Data View window should look like the one displayed on the following page.

---

[27] Indeed following a no-confidence vote that resulted in part from this statement, he resigned as president of Harvard.

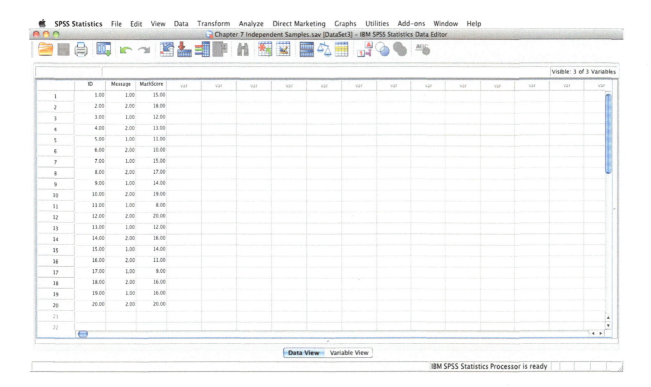

Conducting an Independent-Samples *t* Test
(Analyze→Compare Means→Independent-Samples T Test)

To conduct an independent-samples *t* test, you need to use the upper toolbar to go to **Analyze→Compare Means→Independent-Samples T Test**.

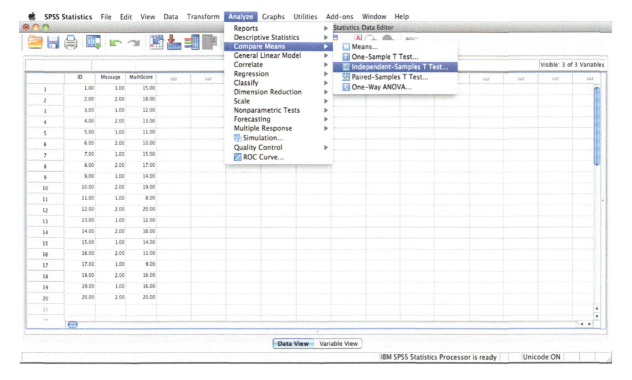

An 'Independent-Samples T Test' dialog window like the one shown below will open. Highlight the dependent variable—**MathScore**—by clicking on the variable name and then click the corresponding **blue arrow** to move it over to the **Test Variable(s) box**. Next, highlight the independent variable—**Message**—by clicking on the variable name, and move it over to the **Grouping Variable box** by clicking the corresponding **blue arrow**. Next, click on the **Define Groups button** (highlighted in the image shown below).

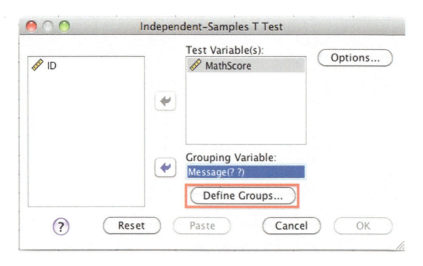

A 'Define Groups' dialog window like the one shown below will now appear. Since we used 1s and 2s to identify the two conditions, you should enter **1** into the box labeled **Group 1** and **2** into the box labeled **Group 2**. Close the dialog window by clicking **Continue**. Finally, click **OK** to close the 'Independent-Samples T Test' dialog window and initiate the analysis.

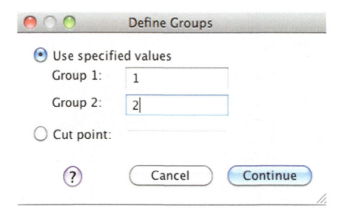

Interpreting the Results

An output window will now appear displaying the following tables. We will first examine the 'Group Statistics' table shown below.

Group Statistics

	Message	N	Mean	Std. Deviation	Std. Error Mean
MathScore	Genetics	10	12.6000	2.67499	.84591
	Socialization	10	16.0000	3.59011	1.13529

This first column of the table displays the dependent variable under consideration (MathScore) and the labels we provided for the two conditions. The column labeled 'N' shows the number of participants in each condition (there were 10 participants in each condition). The column labeled 'Mean' displays the mean math test score in each of the two conditions. You should be able to see that the mean math test score of women in the genetics condition was 12.60 and the mean math test score of women in the socialization condition was 16.00. Note that women's scores were lower in the genetics condition. The column labeled 'Std. Deviation' shows the standard deviation of scores in each condition. Finally, the column labeled 'Std. Error Mean' provides the standard error of the mean for each condition.

The primary results of the analysis are displayed in the 'Independent Samples Test' table shown below.

Independent Samples Test

		Levene's Test for Equality of Variances		t-test for Equality of Means					95% Confidence Interval of the Difference	
		F	Sig.	t	df	Sig. (2-tailed)	Mean Difference	Std. Error Difference	Lower	Upper
MathScore	Equal variances assumed	.604	.447	-2.401	18	.027	-3.40000	1.41578	-6.37445	-.42555
	Equal variances not assumed			-2.401	16.639	.028	-3.40000	1.41578	-6.39199	-.40801

By default, SPSS tests the assumption of homogeneity of variance that underlies the independent-samples *t* test. This is the assumption that the variances of the populations from which the samples were drawn are not significantly different. The section of the table labeled 'Levene's Test for Equality of Variances' contains the results of the test of this assumption. If the results of Levene's test are not significant ($p > .05$), then the variances are not significantly different, meaning the assumption has been met. If the results of Levene's test are significant ($p < .05$), then the variances are significantly different, meaning the data violate the assumption of homogeneity of variance. We do not worry too much about violating this assumption because the independent-samples *t* test is robust to (i.e., affected very little by) violations of this assumption, especially when an equal number of participants are used in each condition. SPSS suggests you report the adjusted values in the bottom row (the row labeled 'Equal variances not assumed') when the assumption has been violated. However, most researchers report the results in the first row regardless of the outcome of this test and simply report when the assumption was violated. You can see by looking at the column labeled 'Sig.' that for our analysis the result of Levene's test was not significant ($p = .45$). Hence, we can conclude that the variances are not significantly different, and as such, that the assumption of homogeneity of variance has been met.

The column labeled 't' displays the obtained *t* value, the column labeled 'df' gives the degrees of freedom, and the column labeled 'Sig. (2-tailed)' provides the *p* value for a two-tailed test. Since the *p* value provided in the table is less than .05, we can conclude that women's math test performance was significantly affected by the message they overheard. The next piece of information we can extract from the table is the difference in the means of the two groups, which is presented in the column labeled 'Mean Difference.' The column labeled 'Std. Error Difference' provides the standard error of the mean for the difference scores (which is the estimated standard deviation of the sampling distribution of the difference between sample means).

Finally, the last section of the table, labeled '95% Confidence Interval of the Difference,' displays the lower and upper limits of the 95% confidence interval for the difference in means. The interval reported in the table indicates that we can be 95% certain that the interval −6.37 to −0.43 contains the true effect of the message that women are genetically predisposed to be poor at math on women's actual math test performance. In other words, we can be 95% certain that exposure to genetic explanations for women's math abilities decreases their math performance by 0.43 to 6.37 points. Since the confidence interval does not cross 0, we can conclude that this effect is statistically significant.

Calculating the Effect Size (Cohen's d)

Once again, SPSS does not contain an option to compute the effect size indicator Cohen's *d*. However, the output does contain the values that we need to calculate it by hand. The formula for computing Cohen's *d* for an independent-samples *t* test is:

$$d = \frac{|M_1 - M_2|}{SD_{\text{Pooled}}}$$

The formula shows that the Cohen's *d* value is computed by dividing the absolute value of the difference in the sample means ($M_1 - M_2$) by the pooled standard deviation (SD_{Pooled}). The pooled standard deviation is essentially a weighted average of the standard deviations of the two groups. It can be computed using the following formula:

$$SD_{\text{Pooled}} = \sqrt{\frac{(n_1 - 1)SD_1^2 + (n_2 - 1)SD_2^2}{n_1 + n_2 - 2}}$$

The values we need to input into these formulas are all provided in the 'Group Statistics' table shown on page 124. As reviewed previously, the mean of condition 1 (M_1) is 12.60, the mean of condition 2 (M_2) is 16.00, the standard deviation of condition 1 (SD_1) is 2.67, and the standard deviation of condition 2 (SD_2) is 3.59. Finally, there were 10 participants in each condition so n_1 and n_2 both equal 10. We need to start by computing the pooled standard deviation:

$$SD_{\text{Pooled}} = \sqrt{\frac{(10-1)2.6750^2 + (10-1)3.5901^2}{10+10-2}} = \sqrt{\frac{(9)7.1556 + (9)12.8888}{18}} = \sqrt{10.2222} = 3.1658$$

We are now ready to find the value of Cohen's *d* by inputting the values of the means and pooled standard deviation into the formula:

$$d = \frac{|12.6000 - 16.0000|}{3.1658} = 1.07$$

Our Cohen's *d* value is 1.07, which indicates that there is a large effect of the message the women overheard on their math test performance. Specifically, it indicates that the women who overhead the bogus message that there is a sex-linked math gene performed 1.07 standard deviations lower on the math test than the women who overheard the message that socialization accounts for women's underrepresentation in the sciences.

Reporting the Results

We can now report the following:

> An independent-samples *t* test revealed that the math test scores of women in the genetics condition ($M = 12.60$, $SD = 2.67$) were significantly lower than the math test scores of women in the socialization condition ($M = 16.00$, $SD = 3.59$), $t(18) = -2.40$, $p = .03$, $d = 1.07$, 95% CI

[−6.37, −0.43]. These results suggest that statements that women have a genetic disadvantage can adversely affect their math test performance.[28]

Calculating the 99% Confidence Intervals for the Difference in Means

Once again, by default SPSS provides the 95% confidence interval for the difference in means. However, this setting can easily be changed to produce any confidence interval you'd like. We'll practice by changing the setting to the 99% confidence interval (which, once again, corresponds to an alpha of .01).

Go to **Analyze→Compare Means→Independent-Samples T Test**. The 'Independent-Samples T Test' dialog window should still show the dependent variable—MathScore—in the Test Variables box, and the independent variable—Message—in the Grouping Variable box. The groups should still be defined as 1 and 2. Simply click the **Options button** in the 'Independent-Samples T Test' dialog window to open the 'Independent-Samples T Test: Options' dialog window shown below. Replace the value of 95 with a value of **99** by simply typing in the box. Press **Continue** to close the dialog window and press **OK** to run the analysis.

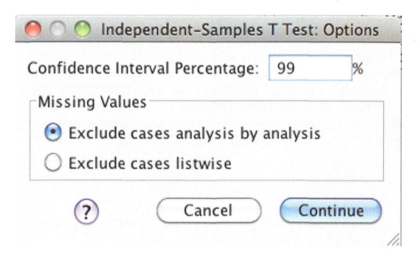

The same tables will appear in the output window. The only difference between these tables and the ones produced previously is highlighted in the 'Independent Samples Test' table shown below.

Independent Samples Test

		Levene's Test for Equality of Variances		t-test for Equality of Means					99% Confidence Interval of the Difference	
		F	Sig.	t	df	Sig. (2-tailed)	Mean Difference	Std. Error Difference	Lower	Upper
MathScore	Equal variances assumed	.604	.447	−2.401	18	.027	−3.40000	1.41578	−7.47525	.67525
	Equal variances not assumed			−2.401	16.639	.028	−3.40000	1.41578	−7.51431	.71431

The table now displays the 99% confidence interval for the difference in means. According to this confidence interval, we can be 99% certain that the interval −7.48 to 0.68 contains the true effect of the statement

[28] While the data presented in this guide are all fabricated, researchers at the University of British Columbia conducted a similar study and found results consistent with these. The reference for the paper is: Dar-Nimrod, I., & Heine, S. J. (2006). Exposure to scientific theories affects women's math performance. *Science*, 314, 435. DOI: 10.1126/science.1131100.

that women are genetically predisposed to be poor at math on women's actual math test performance. Since the confidence interval crosses 0, we must conclude that the effect is not statistically significant. Note that we would have reached this same conclusion using the traditional p value method and an alpha of .01 (the alpha level that corresponds to the 99% confidence interval) because $p = .03 > .01$.

Reporting the Results

The results using an alpha of .01 and the 99% confidence interval would be reported in the following manner:

> Using an alpha of .01, an independent-samples t test revealed that the math test scores of women in the genetics condition ($M = 12.60$, $SD = 2.67$) were not significantly different than the math test scores of women in the socialization condition ($M = 16.00$, $SD = 3.59$), $t(18) = -2.40$, $p = .03$, $d = 1.07$, 99% CI [-7.48, 0.68].

One-Tailed Independent-Samples *t* Test

Now assume you had originally made a directional hypothesis and predicted that the message that women are genetically predisposed to be poor at math would *decrease* their math test performance. In this scenario, you would need to conduct a one-tailed independent-samples t test. The method of conducting the analysis would remain the same, and the tables in the output would remain the same. The only thing that would change in this scenario is the reported p value, which would once again be cut in half. The p value for the two-tailed test was .0273, so the p value for a one-tailed test would be:

$$p = .0273/2$$
$$p = .01$$

Since the mean of the genetics condition was lower than the mean of the socialization condition, we know the results are in the predicted direction.

Reporting the Results

In this scenario of a one-tailed independent-samples t test (with alpha of .05), we would report the following:

> A one-tailed independent-samples t test revealed that the math test scores of women in the genetics condition ($M = 12.60$, $SD = 2.67$) were significantly lower than the math test scores of women in the socialization condition ($M = 16.00$, $SD = 3.59$), $t(18) = -2.40$, $p = .01$ (one-tailed), $d = 1.07$.

Predicting the Wrong Direction

Once again, if the results had worked out in the opposite direction to which we predicted and showed that math test scores were actually higher in the genetics condition, then we would need to make a further adjustment to the p value. Rather than simply dividing the p value in half as we did when the results worked out in the predicted direction, we would now need to subtract the divided p value from a value of 1.

$$p = 1 - (.0339/2)$$
$$p = 1 - .0137$$
$$p = .99$$

In this case, we would report the following: $t(18) = -2.40$, $p = .99$ (one-tailed), $d = 1.07$. Since $p > .05$ in this scenario, we would need to conclude that statements that women have a genetic disadvantage do not adversely affect their math test performance.

One-Way Between Groups ANOVA

Learning Objectives

In this chapter, you will learn how to analyze data using a one-way between groups ANOVA. You will also learn how to calculate and interpret the effect size indicator eta-squared and perform follow-up post hoc tests.

ONE-WAY ANOVA

As reviewed in Chapter 7, *t* tests are used to compare the means of *two* groups. One-way ANOVA is simply an extension of the *t* test; it is used to compare the means of *three or more* groups. We will consider only one-way between groups ANOVA, which is used in conjunction with independent groups (between-subjects) designs in which the various conditions contain different groups of participants. Unfortunately, the limited student version of SPSS does not contain the option to analyze data using one-way within groups ANOVA (which is used in conjunction with repeated measures designs) and as such we will not consider it further.

Assume you learned about the results of the study described in Chapter 7 on the influence of statements about women's math abilities on their math test performance. You decide that you could improve upon the study by including a control condition in which women overhear a message unrelated to math abilities. Prior to collecting any data you decide to set alpha at .05.

You invite 30 women to your lab and you randomly assign each woman to one of three separate conditions. Women in each of the conditions are told to wait in a waiting area where they "accidently overhear" a staged conversation between two professors. Women in condition 1 (the genetics condition) overhear the professors discussing a new paper revealing evidence that a math gene has been discovered that is sex-linked and commonly deficient in women and that this may account for why women are underrepresented in the sciences. Women in condition 2 (the socialization condition) overhear the professors discussing a new paper revealing evidence that women only tend to be underrepresented in the sciences because they are raised in a society where they are regarded as being inferior at math. Women in condition 3 (the control condition) overhear the professors discussing their plans for the weekend. Participants are then asked to complete a math test containing 20 questions. You hypothesize that women's math test performance will be affected by the message they overhear.

Assume you record the following data:

ID	Message	MathScore	ID	Message	MathScore
1	1	12	16	2	20
2	1	11	17	2	20
3	1	15	18	2	15
4	1	15	19	2	20
5	1	14	20	2	19
6	1	14	21	3	18
7	1	12	22	3	9
8	1	8	23	3	14
9	1	9	24	3	15
10	1	16	25	3	9
11	2	20	26	3	11
12	2	15	27	3	14
13	2	11	28	3	15
14	2	20	29	3	15
15	2	17	30	3	20

Begin by **entering the data** into a **blank SPSS spreadsheet**. Note that the top row contains **the variable names**, which must be entered in the **Variable View** window. While you are in Variable View, use the **Values** column to identify message **1** as **Genetics**, message **2** as **Socialization**, and message **3** as **Control**.

Set the scale of **Measure** to **Nominal** for the variables **ID** and **Message**, and to **Scale** for the variable **MathScore**. The remaining data must be entered into three columns and 30 rows in the Data View window. Keep in mind that different variables must be entered into separate columns, while the data for each of the participants must be entered in the rows (therefore each row must represent one participant's data on all of the variables). Once the data are entered, your Data View window should look like the one displayed below.

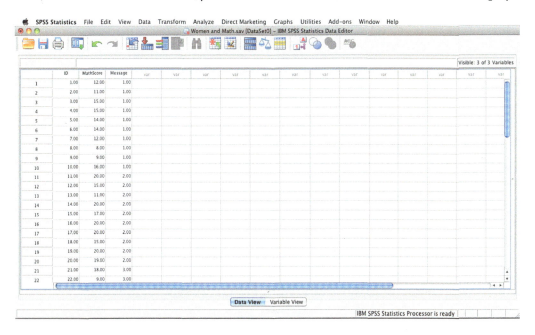

Conducting a One-Way Between Groups ANOVA
(Analyze→General Linear Model→Univariate)

To conduct a one-way between groups ANOVA, you need to use the upper toolbar to go to **Analyze→General Linear Model→Univariate**.[29]

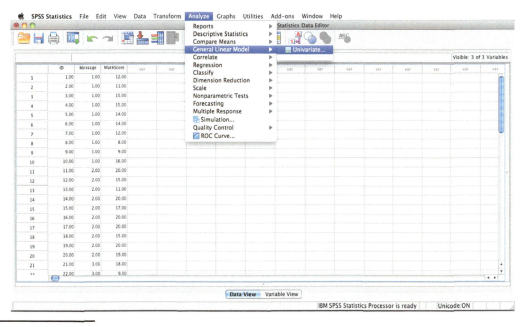

---

[29] The analysis can also be performed using the option Analyze→Compare Means→One-Way ANOVA, however this method does not provide an option to compute an effect size indicator.

Next, a 'Univariate' dialog window like the one shown below will open. Highlight the dependent variable—**MathScore**—by clicking on the variable name, and then click the corresponding **blue arrow** to move it over to the **Dependent Variable box**. Next, highlight the independent variable—**Message**—by clicking on the variable name, and move it over to the **Fixed Factor(s) box** by clicking the corresponding **blue arrow**. Next, click on the **Options button** (highlighted in the image shown below).

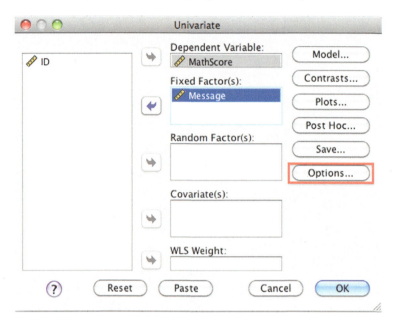

A 'Univariate: Options' dialog window like the one shown below will now appear. Check the options for **Descriptive statistics**, **Estimates of effect size**, and **Homogeneity tests**. Close the dialog window by clicking **Continue**.

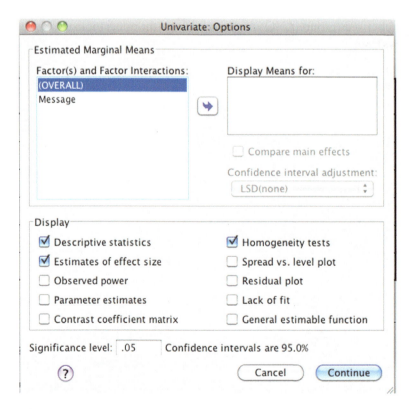

Next, click the **Post Hoc button** (highlighted in the 'Univariate' dialog window shown below).

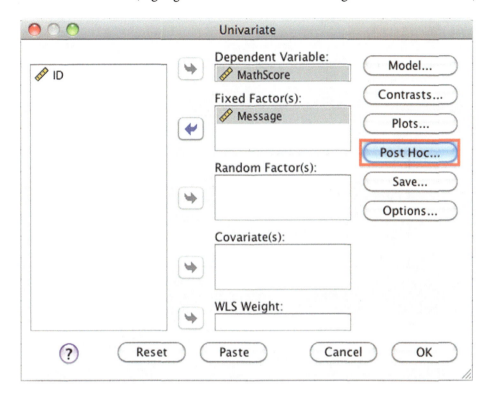

This will open the 'Univariate: Post Hoc Multiple Comparisons for Observed Means' dialog window shown below. Move the independent variable **Message** over to the **Post Hoc Tests for box** by clicking on the blue arrow. **Check** the **Bonferroni** and **Tukey** options, and then click **Continue** to close the dialog window. Finally, click **OK** on the 'Univariate' dialog window to initiate the analysis.

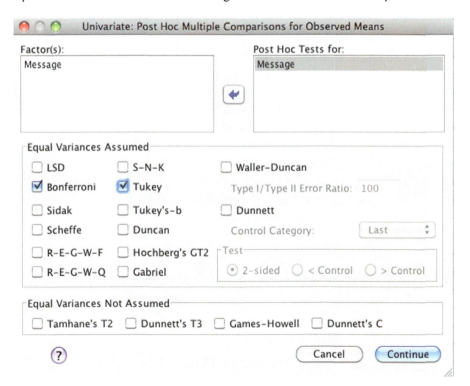

Interpreting the Results

Descriptive Statistics

An output window will now appear displaying a series of tables. We will begin by examining the 'Descriptive Statistics' table shown below.

Descriptive Statistics

Dependent Variable: MathScore

Message	Mean	Std. Deviation	N
Genetics	12.6000	2.67499	10
Socialization	17.7000	3.12872	10
Control	14.0000	3.55903	10
Total	14.7667	3.73874	30

The table displays the labels we entered for the three conditions in the column labeled 'Message,' as well as the means, standard deviations, and sample sizes for each of the three conditions in the columns labeled 'Mean,' 'Std. Deviation,' and 'N,' respectively. In addition, the bottom row labeled 'Total' displays the overall mean, standard deviation, and sample size for all three conditions combined.

Levene's Test of Homogeneity of Variance

As you may recall from Chapter 7, the independent-samples t test carries the assumption of homogeneity of variance (the assumption that the variances of the populations from which the samples were drawn are not significantly different), and this assumption is tested using Levene's test. Since the one-way between groups ANOVA is simply an extension of the independent-samples t test, it also carries this same assumption and it is also tested using Levene's test.

The table labeled 'Levene's Test of Equality of Error Variances,' shown below, displays the results of this test of the assumption of homogeneity of variance. If the p value displayed in the column labeled 'Sig.' is .05 or lower, then the variances are significantly different and the assumption has been violated. This violation should then be reported along with the main results of the ANOVA. If the p value displayed in the table is greater than .05, then the variances are not significantly different and the assumption has been met. As you can see below, for our example the p value associated with Levene's test is .86, which is greater than .05, therefore the assumption of homogeneity of variance has been met.

Levene's Test of Equality of Error Variances[a]

Dependent Variable: MathScore

F	df1	df2	Sig.
.154	2	27	.858

Tests the null hypothesis that the error variance of the dependent variable is equal across groups.

a. Design: Intercept + Message

Main Effect

The primary results of the analysis are displayed in the table labeled 'Tests of Between-Subjects Effects' shown below.

Tests of Between-Subjects Effects

Dependent Variable: MathScore

Source	Type III Sum of Squares	df	Mean Square	F	Sig.	Partial Eta Squared
Corrected Model	138.867[a]	2	69.433	7.035	.003	.343
Intercept	6541.633	1	6541.633	662.755	.000	.961
Message	138.867	2	69.433	7.035	.003	.343
Error	266.500	27	9.870			
Total	6947.000	30				
Corrected Total	405.367	29				

a. R Squared = .343 (Adjusted R Squared = .294)

The column labeled 'Type III Sum of Squares' provides the values of the sums of squares (the sum of the squared deviation scores). The column labeled 'Mean Square' provides the values of the mean squares (estimates of the population variances). The column labeled 'df' contains the degrees of freedom, the column labeled 'F' provides the F statistics, the column labeled 'Sig.' lists the p values, and the column labeled 'Partial Eta Squared' provides the effect size estimates.

The F statistic is the primary statistic considered in ANOVA. It is a ratio of two types of variance; it is a ratio of between groups variance over within groups variance. Between groups variance reflects the variability of the means *across* the groups, and within groups variance reflects the variability of scores *within* each group. Within groups variance is essentially error variance, and as such, SPSS provides the statistics related to it in the row labeled 'Error.' Between groups variance essentially reflects the effect of the independent variable on the dependent variable, and as such, the statistics related to it are provided in the row labeled with the name of the independent variable. We named our independent variable message, therefore the row labeled 'Message' in the table shown above indicates that the between groups sum of squares is 138.87, the between groups degrees of freedom are 2, and the mean square between is 69.43. The row labeled 'Error' shows that the within groups sum of squares is 266.50, the within groups degrees of freedom are 27, and the mean square within is 9.87.

The main effect is the overall effect of the independent variable on the dependent variable. Once again, between groups variance essentially reflects the effect of the independent variable on the dependent variable, and as such, the statistics related to it and to the main effect are provided in the row labeled with the independent variable. Therefore, the primary statistics of interest to us are provided in the row labeled 'Message.' As displayed in that row, the value of F is 7.03, the p value is .003, and the effect size is .34. As alluded to above, there are two degrees of freedom of interest that are associated with the F test, between groups degrees of freedom and within groups degrees of freedom. The value of the between groups degrees of freedom is listed in the row labeled with the name of the independent variable, while the value of the within groups degree of freedom is listed in the row labeled 'Error.' The degrees of freedom are reported in parentheses, separated by a comma.

Putting it all together, the main effect of the message women overheard on their math test scores would be reported as, $F(2, 27) = 7.03$, $p = .003$, $\eta^2 = .34$. These results suggest that there is a significant effect of the message women overheard on their math test performance.

Effect Size (Eta-Squared)

The most commonly used effect size indicator for ANOVA is eta-squared, symbolized as η^2. Since it is a Greek character, it should not be italicized. Moreover, since eta-squared is a proportion whose values cannot exceed 1, leading 0s before the decimal are not appropriate.

Eta-squared values of .01 are considered small, values of .06 are considered medium, and values of .14 are considered large. Eta-squared is also sometimes referred to as R^2 because, like R^2, it indicates the proportion of variability in the dependent variable that is accounted for by the independent variable. Indeed, in addition to displaying the eta-squared value for the main effect of the independent variable in the column labeled 'Partial Eta Squared,'[30] the value is also provided under the table with the label 'R Squared.'

As shown both in and under the table, the eta-squared value of interest to us is .34. This is a large effect and indicates that the message women overheard accounts for 34.26% of the variability in their math test scores.

Post Hoc Tests

A significant main effect (F statistic) indicates that at least one of the means differs from at least one of the other means. However, the F statistic does not provide any indication of which means differ significantly from which. In order to determine which means differ significantly from each other, we need to use post hoc tests. Thus, post hoc tests are only necessary (and in most cases only appropriate) when the overall main effect is statistically significant. This is because if the F statistic is not significant then it indicates that none of the means differ significantly from one another and any follow-up analysis using post hoc tests could been seen as a fishing exhibition. Since our main effect was statistically significant, we will need to consider the results of the post hoc tests to determine which means differ significantly. The results of the post hoc comparisons are displayed in the table shown below.

Multiple Comparisons

Dependent Variable: MathScore

	(I) Message	(J) Message	Mean Difference (I–J)	Std. Error	Sig.	95% Confidence Interval Lower Bound	95% Confidence Interval Upper Bound
Tukey HSD	Genetics	Socialization	−5.1000*	1.40502	.003	−8.5836	−1.6164
		Control	−1.4000	1.40502	.585	−4.8836	2.0836
	Socialization	Genetics	5.1000*	1.40502	.003	1.6164	8.5836
		Control	3.7000*	1.40502	.036	.2164	7.1836
	Control	Genetics	1.4000	1.40502	.585	−2.0836	4.8836
		Socialization	−3.7000*	1.40502	.036	−7.1836	−.2164
Bonferroni	Genetics	Socialization	−5.1000*	1.40502	.004	−8.6862	−1.5138
		Control	−1.4000	1.40502	.984	−4.9862	2.1862
	Socialization	Genetics	5.1000*	1.40502	.004	1.5138	8.6862
		Control	3.7000*	1.40502	.041	.1138	7.2862
	Control	Genetics	1.4000	1.40502	.984	−2.1862	4.9862
		Socialization	−3.7000*	1.40502	.041	−7.2862	−.1138

Based on observed means.
   The error term is Mean Square(Error) = 9.870.

   *. The mean difference is significant at the

---

[30] For one-way ANOVA, partial eta-squared is equal to eta-squared, but since eta-squared is a more informative and commonly used statistic, we will report the statistic as an eta-squared value rather than a partial eta-squared value. A more complete explanation of the difference between eta-squared and partial eta-squared is provided in Chapter 9.

The top section of the table, labeled 'Tukey HSD,' contains the results of Tukey's Honestly Significant Difference (*HSD*) post hoc test. The bottom section of the table, labeled 'Bonferroni,' contains the results of the Bonferroni post hoc test. Both are highly respected and commonly used post hoc tests that control for the inflation in Type I error that could occur with multiple comparisons (i.e., they keep the overall alpha level for the complete series of tests at .05). The Bonferroni test tends to be a bit more conservative than Tukey's *HSD* test. Typically, only one post hoc analysis is performed and reported. Two types are reported here only to offer a description and demonstration of each of these commonly used tests.

The columns labeled '(I) Message' and '(J) Message' display the conditions being compared. The column labeled 'Mean Difference (I − J)' shows the differences in the means of the two conditions being compared. The column labeled 'Std. Error' contains the standard error of the means. The column labeled 'Sig.' contains the *p* values for the comparisons, and the final columns contain the 95% confidence intervals for the difference in the means.

Let's start with the results of Tukey's *HSD* test shown in the upper portion of the table. The first row contains a comparison of the genetics and socialization conditions. The value of −5.10 in the column labeled 'Mean Difference (I − J)' indicates that the mean of the genetics condition is 5.10 units lower than the mean of the socialization condition. The *p* value of .003 displayed in the 'Sig.' column indicates that this difference is statistically significant. The next row contains the comparison of the genetics and control conditions. The *p* value of .59 listed in this row indicates that the difference (of −1.40) in the means of these two conditions is not statistically significant. The next row, comparing the socialization and genetics conditions, is redundant with the first row. However, the following row, comparing the socialization and control conditions, reveals that the difference of 3.70 in the means of these two conditions is statistically significant (*p* = 04). The final two rows of the upper portion of the table are redundant with the second and fourth rows that we have already reviewed.

A review of the bottom portion of the 'Multiple Comparisons' table reveals that the results of the comparisons using the Bonferroni procedure are consistent with those of Tukey's *HSD* test. Specifically, they show that the mean difference of −5.10 across the genetics and socialization conditions is statistically significant (*p* = .004), that the difference of −1.40 in the means of the genetics and control conditions is not statistically significant (*p* = .98), and that the mean difference of 3.70 across the socialization and control conditions is statistically significant (*p* = .04).

Reporting the Results

Once again, we would typically only compute and report the results of one post hoc test. Thus, the results of the one-way ANOVA with the Tukey's *HSD* post hoc test are reported as:

> A one-way between groups ANOVA revealed a significant main effect of the message women overheard on their math test performance, $F(2, 27) = 7.03$, $p = .003$, $\eta^2 = .34$. Post hoc comparisons using Tukey's *HSD* test indicated that the math test scores of women in the genetics condition ($M = 12.60$, $SD = 2.67$) were significantly lower than the math test scores of women in the socialization condition ($M = 17.70$, $SD = 3.13$) ($p = .003$). Women in the control condition also had significantly lower math test scores ($M = 14.00$, $SD = 3.56$) than those in the socialization condition ($p = .04$). The difference between the math test scores of women in the genetics and control conditions was not significant ($p = .59$).

9

Between Groups Factorial ANOVA

Learning Objectives

In this chapter, you will learn how to analyze data using between groups factorial ANOVA. You will learn how to calculate and interpret the effect size indicators eta-squared and partial eta-squared. Complete examples of a 2 × 2 and a 2 × 3 between groups factorial ANOVA will demonstrate when and how to perform simple main effects analyses and follow-up post hoc comparisons.

FACTORIAL ANOVA

As described in Chapter 8, one-way ANOVA is used to examine the effect of *one* independent variable on the means of three or more groups. Factorial ANOVA is just an extension of this; it is used to examine the effects of *more than one* independent variable on the means of two or more groups.

We will consider only between groups factorial ANOVA, which is used in conjunction with independent groups (between-subjects) designs. This is because the limited student version of SPSS does not contain the options to analyze data using within groups factorial ANOVA (which is used in conjunction with repeated measures designs) or mixed factorial ANOVA (which is used in conjunction with mixed designs).

We will start by considering a 2 × 2 between groups factorial ANOVA, which is used to analyze the data from experiments with two independent variables, each of which has two levels (i.e., two conditions). We will then proceed to a 2 × 3 between groups factorial ANOVA, which is used to analyze the data from experiments with two independent variables, one of which has two levels and the other of which has three levels.

2 × 2 BETWEEN GROUPS FACTORIAL ANOVA

Assume you learned about the results of the experiment described in Chapter 7 on the influence of statements about women's math abilities on their math performance. You decide that you could improve upon the study by including control conditions in which men also overhear the two messages related to women's math abilities. You hypothesize that women's math test performance will be affected by the message they overhear, while men's math test performance will be unaffected by these messages. Prior to collecting any data you decide to set alpha at .05.

You invite 20 women and 20 men to your lab and you randomly assign each individual to one of two separate conditions in such a way that there are 10 men and 10 women in each condition. Participants in each of the conditions are asked to wait in a waiting area where they "accidently overhear" a staged conversation between two professors. Participants in condition 1 (the genetics condition) overhear the professors discussing a new paper revealing evidence that a math gene has been discovered that is sex-linked and commonly deficient in women and that this may account for why women are underrepresented in the sciences. Participants in condition 2 (the socialization condition) overhear the professors discussing a new paper revealing evidence that women only tend to be underrepresented in the sciences because they are raised in a society where they are regarded as being inferior at math.

Since there are two independent variables (message and gender), each with two levels (genetics/socialization and male/female), and each of the conditions contain different people, this would be considered a 2 × 2 between groups factorial design. Accordingly, the data will need to be analyzed using a 2 × 2 between groups factorial ANOVA.

Assume you record the following data:

ID	Message	Gender	MathScore	ID	Message	Gender	MathScore
1	1	0	9	21	1	1	17
2	1	0	12	22	1	1	19
3	1	0	14	23	1	1	12
4	1	0	16	24	1	1	20
5	1	0	11	25	1	1	18
6	1	0	12	26	1	1	17
7	1	0	14	27	1	1	10
8	1	0	7	28	1	1	14
9	1	0	10	29	1	1	16
10	1	0	17	30	1	1	12
11	2	0	19	31	2	1	11
12	2	0	14	32	2	1	14
13	2	0	12	33	2	1	15
14	2	0	20	34	2	1	15
15	2	0	18	35	2	1	20
16	2	0	18	36	2	1	11
17	2	0	14	37	2	1	14
18	2	0	12	38	2	1	15
19	2	0	19	39	2	1	15
20	2	0	18	40	2	1	20

Begin by **entering the data** into a **blank SPSS spreadsheet**. Note that the top row contains the **variable names**, which must be entered in the **Variable View** window. While you are in Variable View, use the **Values** column to identify message 1 as **Genetics** and message 2 as **Socialization**. Also use the **Values** column to label gender **0 as Female** and gender **1 as Male**. Set the scale of **Measure** to **Nominal** for the variables **ID**, **Gender**, and **Message**, and to **Scale** for the variable **MathScore**. The remaining data must be entered into four columns and 40 rows in the Data View window. Once the data are entered, your Data View window should look like the one displayed below. Save this dataset, as we will use it again for the next demonstration.

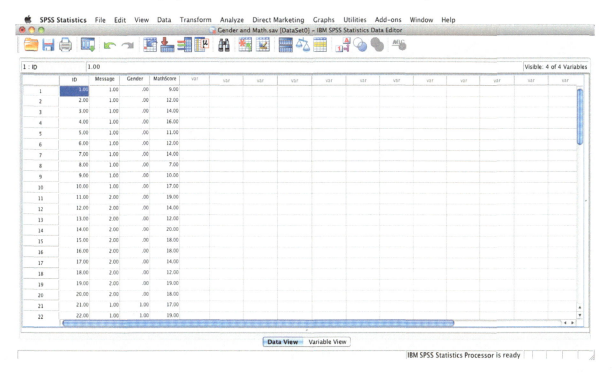

Conducting a Between Groups Factorial ANOVA
(Analyze→General Linear Model→Univariate)

To conduct the between groups factorial ANOVA, you need to use the upper toolbar to go to **Analyze→General Linear Model→Univariate**.

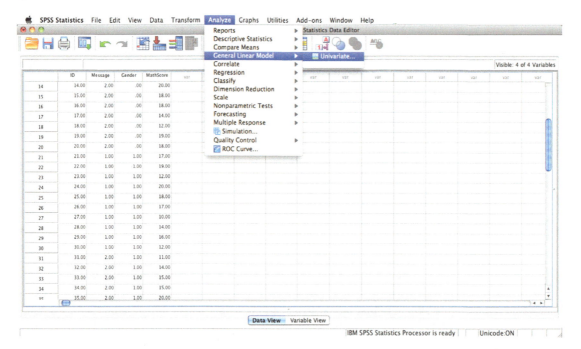

A 'Univariate' dialog window like the one shown below will open. Highlight the dependent variable—**MathScore**—by clicking on the variable name and then click on the corresponding **blue arrow** to move it over to the **Dependent Variable box**. Next, highlight the independent variables—**Message** and **Gender**—by clicking on the variable names, and move them over to the **Fixed Factor(s) box** by clicking the corresponding **blue arrow**. Next, click on the **Options button** (highlighted in the image shown below).

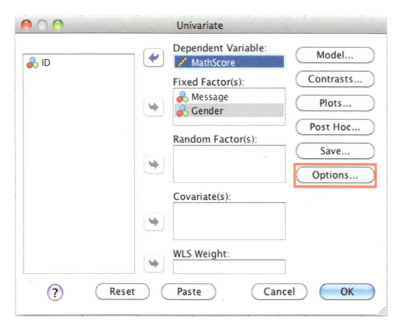

A 'Univariate: Options' dialog window like the one shown below will now appear. Check the options for **Descriptive statistics**, **Estimates of effect size**, and **Homogeneity tests**. Close the dialog window by clicking **Continue**. Finally, click **OK** to close the 'Univariate' dialog window and execute the analysis.

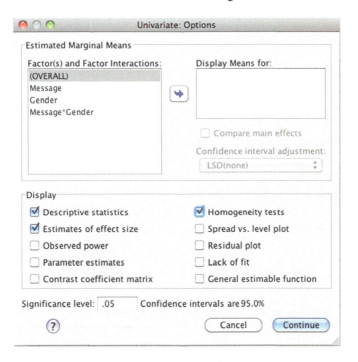

Interpreting the Results

Descriptive Statistics

An output window will now appear displaying a series of tables. We will begin by examining the 'Descriptive Statistics' table shown below.

Descriptive Statistics

Dependent Variable: MathScore

Message	Gender	Mean	Std. Deviation	N
Genetics	Female	12.2000	3.11983	10
	Male	15.5000	3.34166	10
	Total	13.8500	3.57292	20
Socialization	Female	16.4000	3.06232	10
	Male	15.0000	3.05505	10
	Total	15.7000	3.06251	20
Total	Female	14.3000	3.70064	20
	Male	15.2500	3.12671	20
	Total	14.7750	3.41556	40

The table is divided into three main sections. The upper section of the table, labeled 'Genetics,' displays basic descriptive statistics for the genetics condition. Specifically, it displays the means, standard deviations, and sample sizes for the groups of females and males in the genetics condition, as well as for the combined sample of females and males in this condition (in the row labeled 'Total'). The middle section of the table, labeled 'Socialization,' displays the means, standard deviations, and sample sizes for the groups of females and males in the socialization condition, as well as for the combined sample of females and males in this condition. The bottom section of the table, labeled 'Total,' displays the overall means, standard deviations, and sample sizes for females and males in the genetics and socialization conditions combined, as well as for all four groups combined.

Levene's Test of Homogeneity of Variance

The table labeled 'Levene's Test of Equality of Error Variances' shown below displays the results of the test of the assumption of homogeneity of variance. You should recall from previous chapters (Chapters 7 and 8) that if the p value displayed in the column labeled 'Sig.' is .05 or lower, then the variances are significantly different and the assumption has been violated. This violation should then be reported along with the main results of the factorial ANOVA. If the p value displayed in the table is greater than .05, then the variances are not significantly different and the assumption has been met. As you can see below, the p value associated with Levene's test is .72, which is greater than .05, therefore the assumption of homogeneity of variance has been met.

Levene's Test of Equality of Error Variances[a]

Dependent Variable: MathScore

F	df1	df2	Sig.
.449	3	36	.719

Tests the null hypothesis that the error variance of the dependent variable is equal across groups.

a. Design: Intercept + Message + Gender + Message * Gender

Main Effects

The primary results of the analysis are displayed in the table labeled 'Tests of Between-Subjects Effects' shown below.

Tests of Between-Subjects Effects

Dependent Variable: MathScore

Source	Type III Sum of Squares	df	Mean Square	F	Sig.	Partial Eta Squared
Corrected Model	98.475[a]	3	32.825	3.315	.031	.216
Intercept	8732.025	1	8732.025	881.775	.000	.961
Message	34.225	1	34.225	3.456	.071	.088
Gender	9.025	1	9.025	.911	.346	.025
Message * Gender	55.225	1	55.225	5.577	.024	.134
Error	356.500	36	9.903			
Total	9187.000	40				
Corrected Total	454.975	39				

a. R Squared = .216 (Adjusted R Squared = .151)

As described in Chapter 8, the column labeled 'Type III Sum of Squares' provides the values of the sums of squares (the sum of the squared deviation scores). The column labeled 'Mean Square' provides the values of the mean squares (estimates of the population variances). The column labeled 'df' contains the degrees of freedom, the column labeled 'F' provides the F statistics, the column labeled 'Sig.' lists the p values, and the column labeled 'Partial Eta Squared' provides the partial eta-squared values (effect size estimates).

For one-way ANOVA, there is only one independent variable, and thus only the main effect of that single independent variable is examined and reported. In contrast, for a 2×2 factorial ANOVA, there are two independent variables, and as such, two main effects must be considered. Each main effect reflects the overall effect of one independent variable on the dependent variable. Most of the statistics relevant to each main effect are displayed in the row labeled with the name of the independent variable.

We named the first independent variable 'Message,' and therefore the statistics related to the main effect of the message participants overheard are provided in the row labeled 'Message.' As displayed in that row, the value of F is 3.46, the p value is .07, and the partial eta-squared value is .09. Recall from Chapter 8 that there are two degrees of freedom of interest associated with the F test. The between groups degrees of freedom is listed in the row labeled with the name of the independent variable, while the within groups degrees of freedom is listed in the row labeled 'Error.' Putting it all together, the main effect of the message participants overheard on their math test scores would be reported as, $F(1, 36) = 3.46$, $p = .07$, $\eta_p^2 = .09$.

We named the second independent variable 'Gender,' and therefore the statistics related to the main effect of gender are provided in the row labeled 'Gender.' As displayed in that row, the value of F is 0.91, the p value is .35, and the partial eta-squared value is .02. The between groups degrees of freedom is 1, and the within groups degrees of freedom is 36. Putting it all together, the main effect of gender on math test scores would be reported as, $F(1, 36) = 0.91$, $p = .35$, $\eta_p^2 = .02$.

Interaction

Factorial ANOVA also differs from one-way ANOVA in that it considers the interaction between the independent variables. In other words, factorial ANOVA allows us to determine whether the effect of one independent variable on the dependent variable *depends* on the other independent variable. The statistics related to the interaction between the independent variables are provided in the row labeled with a star between the names of the independent variables. In our case, they are displayed in the row labeled 'Message * Gender.' As displayed in that row, the value of F is 5.58, the p value is .02, and the partial eta-squared value is .13. The between groups degrees of freedom is 1, and the within groups degrees of freedom is 36. Putting it all together, the interaction between gender and the message overheard on math test scores would be reported as, $F(1, 36) = 5.58$, $p = .02$, $\eta_p^2 = .13$.

The presence of an interaction indicates that the main effects are misleading. Main effects reflect the *overall* effect of the independent variable on the dependent variable, *ignoring* the effects of the other independent variable (collapsing across the levels of the other independent variable). The presence of a significant interaction indicates that the effects of the independent variables *depend* on the level of the other independent variable, and therefore it is not appropriate to collapse across or ignore that other independent variable. Concretely, for this example, the absence of significant main effects of gender and message suggest that neither gender nor the message overheard has a significant influence on math test scores. However, the interaction indicates that the effects of the message overheard on math test scores depends on gender and/or that the effects of gender on math test scores depends on the message overheard.

Effect Sizes—Eta-Squared (η^2) versus Partial Eta-Squared (η_p^2)

Once again, the most commonly used effect size indicator for ANOVA is eta-squared, symbolized η^2. However, SPSS only computes and reports a related but different statistic called partial eta-squared, which is symbolized as η_p^2. Remember, eta is a Greek character, so it should not be italicized. Moreover, since the value cannot exceed 1, leading 0s before the decimal are not appropriate.

The difference between eta-squared and partial eta-squared is analogous to the difference between the coefficient of determination (r^2) and the squared partial correlation ($r^2_{ab.c}$). You may recall that, while the coefficient of determination is an indicator of the proportion of variability in the criterion (Y) accounted for by the predictor (X), the squared partial correlation is an indicator of the proportion of variability in the criterion (Y) accounted for by the predictor (X_1) after the influence of the other predictor variables (X_2, etc.) on the criterion (Y) and the predictor variable of interest (X_1) have been removed. Similarly, eta-squared is an indicator of the proportion of variability in the dependent variable accounted for by the independent variable, while partial eta-squared is an indicator of the variability in the dependent variable accounted for by the independent variable with variability accounted for by the other independent variable(s) and their interactions(s) partialled out.

Eta-squared is calculated by dividing the sum of squares between by the total sum of squares, while partial eta-squared is calculated by dividing the sum of squares between by the sum of the sum of squares between and the sum of squares within.

$$\eta^2 = \frac{SS_B}{SS_{Total}} \quad \eta_P^2 = \frac{SS_B}{SS_B + SS_W}$$

For one-way ANOVA, the total sum of squares is equal to the sum of the sum of squares between and the sum of squares within ($SS_{Total} = SS_B + SS_W$), therefore the values of eta-squared and partial eta-squared will be equal. However, this is not the case for factorial ANOVA. In factorial ANOVA, the total sum of squares also includes the sum of squares for the other independent variables and the sum of squares for the interaction. Therefore, for factorial ANOVA, the values of eta-squared and partial eta-squared will differ.

Since the value of the denominator is generally lower for partial eta-squared, these values will be larger than eta-squared values. Thus, while eta-squared values of .01, .06, and .14 are considered small, medium, and large, respectively; partial eta-squared values of .10, .25, and .50 are considered small, medium, and large, respectively.

Since SPSS only provides the partial eta-squared values, we need to calculate the eta-squared values by hand using the values reported in the 'Tests of Between-Subjects Effects' table and the following formula:

$$\eta^2 = \frac{SS_B}{SS_{Total}}$$

Once again, the sums of squares are displayed in the column labeled 'Type III Sum of Squares.' The sum of squares between (SS_B) for each independent variable is displayed in the row labeled with the name of the independent variable, and the total sum of squares (SS_{Total}) is displayed in the row labeled 'Corrected Total.' Thus, the sum of squares between for the independent variable message is 34.23, the sum of squares between for the independent variable gender is 9.03, and the sum of squares between for the interaction is 55.23. The total sum of squares (which is the sum of these three sums of squares and the sum of squares within) is 454.98. Using these values (rounded to four decimal places), we can now compute the eta-squared values for each of the independent variables and their interaction.

The eta-squared value for the independent variable message is:

$$\eta^2 = \frac{SS_B}{SS_{Total}} = \frac{34.2250}{454.9750} = .08$$

The eta-squared value for the independent variable gender is:

$$\eta^2 = \frac{SS_B}{SS_{Total}} = \frac{9.0250}{454.9750} = .02$$

The eta-squared value for the interaction is:

$$\eta^2 = \frac{SS_B}{SS_{Total}} = \frac{55.2250}{454.9750} = .12$$

These values indicate that there is a medium-sized effect of message on math test scores, a small effect of gender on math test scores, and a fairly large effect of the interaction between message and gender on math test scores.

Simple Main Effects

If the interaction between the independent variables is not statistically significant, then the main effects tell the complete story and no further analyses are required. However, whenever an interaction is present, it suggests that the main effects are misleading, and it is therefore necessary to conduct further analyses to understand the nature of the interaction. These analyses are commonly referred to as simple main effects (or simple effects) analyses. Simple main effects analyses involve examining the effect of each independent variable at each level of the other independent variable. So, while *main effects* analyses allows us to examine the *overall* effect of each independent variable (collapsing across and therefore ignoring the levels of the other independent variable), *simple main effects* analyses allow us to examine the effect of each independent variable *at each level* of the other independent variable. In doing so, simple main effects analyses break down the interaction. Once again, for our example, while the absence of significant main effects of gender and message may at first glance suggest that these variables have no influence on math test scores, the presence of a significant interaction tells us that the influence of message on math test scores varies with gender and/or that the influence of gender on math test scores varies with message.

For a 2 × 2 between groups ANOVA, the simple main effects analysis can be conducted by performing a series of independent-samples t tests.[31] Specifically, we can use four independent-samples t tests to examine the following:

1. The influence of the message on women only
2. The influence of the message on men only
3. The influence of gender in the genetics condition
4. The influence of gender in the socialization condition

Since we are going to use four separate tests, we would be wise to use a Bonferroni correction to alpha in order to reduce our family-wise Type I error rate. If we want to keep our overall alpha level at .05 and we are running four tests, we can simply divide our alpha of .05 by 4 (.05/4 = .0125 = .01) and use the Bonferroni adjusted alpha of .01 to determine statistical significance.

---

[31] We will have to use this quick and dirty method for our simple main effects analysis since the limited student version of SPSS does not contain the necessary syntax editor required to use the more conventional method.

Conducting a Simple Main Effects Analysis for a 2 × 2 Between Groups ANOVA

The Influence of Message (Separated by Gender)

The main effects analysis indicated that the message did not have a significant effect on the combined groups of men and women, but the presence of an interaction once again suggests that the effect of the message differs across the genders. Therefore, we will begin our simple main effects analysis by examining the influence of the message on women and men separately. Since we want to consider these two groups separately, we will need to use the split file function to split our sample into males and females. As reviewed in Chapter 1, the split file function can be found in the upper toolbar under **Data→Split File**.

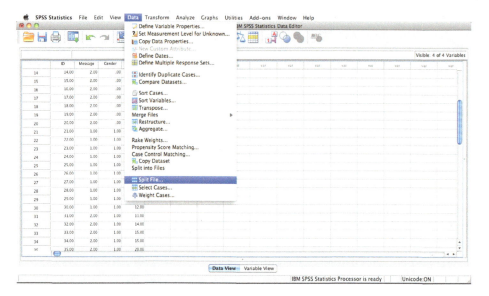

A 'Split File' dialog window will open. Click **Organize output by groups**, and then move the variable **Gender** into the **Groups Based on box**. Click **OK**.

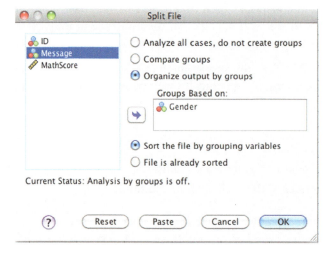

Now that we have split the file into males and females, we can use independent-samples *t* tests to separately compare (1) the women's math test scores in the genetics and socialization conditions, and (2) the men's math test scores in the genetics and socialization conditions. As described in Chapter 7, to perform an independent-samples *t* test, you need to go to **Analyze→Compare Means→Independent-Samples T Test**.

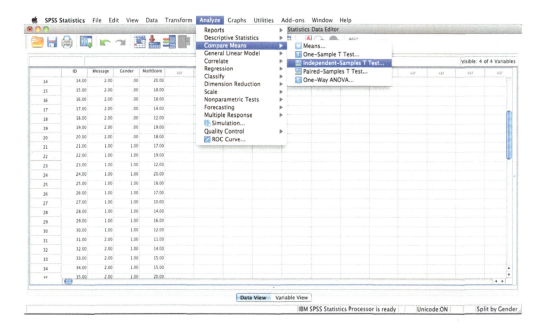

The 'Independent-Samples T Test' dialog window shown below will now open. Move the dependent variable—**MathScore**—into the **Test Variable(s) box**. Since we want to compare math test scores across the genetics and socialization conditions, we need to put the independent variable **Message** into the **Grouping Variable box**. Next, click on the **Define Groups button**.

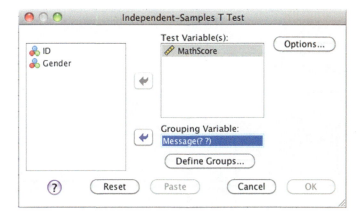

The 'Define Groups' dialog window will now appear. Since we labeled the genetics condition 1 and socialization condition 2, we will need to enter **1** in the **Group 1 box**, and **2** in the **Group 2 box**. Click **Continue**.

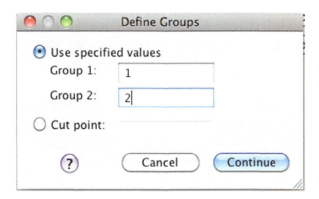

Finally, we need to change the confidence interval setting to the 99% confidence interval to be consistent with our Bonferroni adjusted alpha of .01. Click on the **Options button** in the 'Independent-Samples T Test' dialog window. Using the 'Independent-Samples T Test: Options' dialog window shown below, change the 95% confidence interval default setting to 99% by typing **99** into the **Confidence Interval Percentage box**. Click **Continue** and then **OK** to close the dialog windows and execute the analyses.

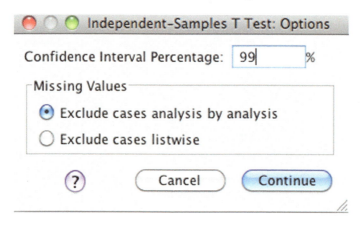

Interpreting the Results

The results of the independent-samples *t* tests will now appear in the output window separated by gender. Let's start by considering the results for the women, shown under the heading 'Gender = Female.'

Gender = Female

Group Statistics[a]

Message		N	Mean	Std. Deviation	Std. Error Mean
MathScore	Genetics	10	12.2000	3.11983	.98658
	Socialization	10	16.4000	3.06232	.96839

a. Gender = Female

Independent Samples Test[a]

		Levene's Test for Equality of Variances		t-test for Equality of Means						
		F	Sig.	t	df	Sig. (2-tailed)	Mean Difference	Std. Error Difference	99% Confidence Interval of the Difference	
									Lower	Upper
MathScore	Equal variances assumed	.183	.674	-3.038	18	.007	-4.20000	1.38243	-8.17924	-.22076
	Equal variances not assumed			-3.038	17.994	.007	-4.20000	1.38243	-8.17940	-.22060

a. Gender = Female

As described completely in Chapter 7, the primary results of interest are displayed in the 'Independent Samples Test' table. As shown in the column labeled 'Sig. (2-tailed),' the *p* value is less than our Bonferroni adjusted alpha of .01, and therefore this comparison is statistically significant. That is, the math test scores of women in the genetics condition ($M = 12.20$, $SD = 3.12$) are significantly lower than the math test scores of women in the socialization condition ($M = 16.40$, $SD = 3.06$), $t(18) = -3.04$, $p = .007$, $d = 1.36$, 99% CI [-8.18, -0.22].

As reviewed in Chapter 7, the Cohen's *d* value reported above needs to be hand calculated using the values shown in the 'Group Statistics' table and the formulas shown below.

$$d = \frac{|M_1 - M_2|}{SD_{Pooled}} \qquad SD_{Pooled} = \sqrt{\frac{(n_1 - 1)SD_1^2 + (n_2 - 1)SD_2^2}{n_1 + n_2 - 2}}$$

The table shows that $n_1 = 10$, $SD_1 = 3.12$, $n_2 = 10$, and $SD_2 = 3.06$. This is all of the information we need to solve for the pooled standard deviation.

$$SD_{Pooled} = \sqrt{\frac{(10-1)3.1198^2 + (10-1)3.0623^2}{10+10-2}} = \sqrt{\frac{(9)9.7332 + (9)9.3777}{18}} = \sqrt{9.5555} = 3.0912$$

The 'Group Statistics' table shows that $M_1 = 12.20$, $M_2 = 16.40$, and we just determined that SD_{Pooled} is 3.09, so we can now solve for Cohen's d.

$$d = \frac{|12.2000 - 16.4000|}{3.0912} = 1.36$$

This Cohen's d value of 1.36 indicates that there is a large effect of message on women's math test scores. Specifically, it indicates that for women, the mean of the genetics condition is 1.36 standard deviation units lower than the mean of the socialization condition.

We will now consider the results for the men shown below.

Gender = Male

Group Statistics[a]

	Message	N	Mean	Std. Deviation	Std. Error Mean
MathScore	Genetics	10	15.5000	3.34166	1.05672
	Socialization	10	15.0000	3.05505	.96609

a. Gender = Male

Independent Samples Test[a]

		Levene's Test for Equality of Variances		t-test for Equality of Means						99% Confidence Interval of the Difference	
		F	Sig.	t	df	Sig. (2-tailed)	Mean Difference	Std. Error Difference	Lower	Upper
MathScore	Equal variances assumed	.871	.363	.349	18	.731	.50000	1.43178	-3.62130	4.62130
	Equal variances not assumed			.349	17.857	.731	.50000	1.43178	-3.62513	4.62513

a. Gender = Male

The tables show that the math test scores of men in the genetics condition ($M = 15.50$, $SD = 3.34$) do not differ significantly from the math test scores of men in the socialization condition ($M = 15.00$, $SD = 3.06$), $t(18) = 0.35$, $p = .73$, $d = 0.16$, 99% CI [−3.62, 4.62].

The Cohen's d value reported above needs to be hand calculated. Using the values in the table, we can solve for the pooled standard deviation.

$$SD_{Pooled} = \sqrt{\frac{(10-1)3.3417^2 + (10-1)3.0551^2}{10+10-2}} = \sqrt{\frac{(9)11.1670 + (9)9.3336}{18}} = \sqrt{10.2503} = 3.2016$$

Using the pooled standard deviation and the means reported in the 'Group Statistics' table, we can now solve for Cohen's d.

$$d = \frac{|15.5000 - 15.0000|}{3.2016} = 0.16$$

This is a small effect and indicates that, for men, the mean of the genetics condition is 0.16 standard deviation units higher than the mean of the socialization condition.

Now we know that one source of the interaction is that the effects of the message depend on gender. The message overheard influenced women's math test scores, but it did not influence men's math test scores. Next, we need to determine whether the effects of gender depend on message. That is, do the effects of gender on math test performance depend on the message that men and women overhear?

The Influence of Gender (Separated by Message)

The main effects analysis indicated that overall the math test performance of men and women did not differ. However, the presence of a significant interaction once again suggests that the effect of gender may differ depending on the message. Therefore, we will continue our simple main effects analysis by examining the influence of gender in the genetics and socialization conditions separately. Since we want to consider these two conditions separately, we will need to use the split file function to split our sample into the genetics and socialization conditions. Go to **Data→Split File**. Using the 'Split File' dialog window shown below, click **Organize output by groups**, and then move the variable **Gender** *out* of the **Groups Based on box**, and move the variable **Message** *into* the **Groups Based on box**. Click **OK**.

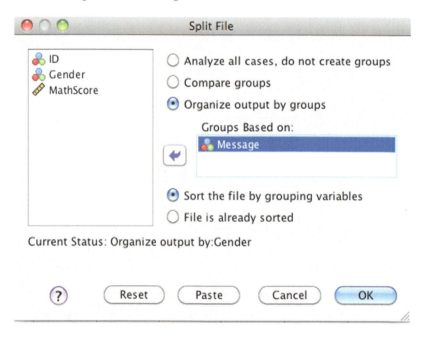

Now that we have split the file into the genetics and socialization conditions, we can use independent-samples *t* tests to compare the women and men's math test performance separately for each condition. To perform these analyses, go to **Analyze→Compare Means→Independent-Samples T Test**. The dependent variable **MathScore** should still appear in the **Test Variable(s) box**, and the independent variable **Message** should still appear in the **Grouping Variable box**. Remove **Message** from the **Grouping Variable box** by clicking on the **blue arrow**. Next, add **Gender** to the **Grouping Variable box**. Click **Define Groups**. Since we coded the genders with 0s and 1s, we will need to enter **0** into the **Group 1 box** and **1** into the **Group 2 box**. Click **Continue** and then **OK** to close both dialog windows and execute the analysis.

Interpreting the Results

The following tables will now appear in the output window. We will start by considering the results for the genetics condition. These results are provided in the output window under the heading 'Message = Genetics.'

Message = Genetics

Group Statistics[a]

	Gender	N	Mean	Std. Deviation	Std. Error Mean
MathScore	Female	10	12.2000	3.11983	.98658
	Male	10	15.5000	3.34166	1.05672

a. Message = Genetics

Independent Samples Test[a]

		Levene's Test for Equality of Variances		t-test for Equality of Means					99% Confidence Interval of the Difference	
		F	Sig.	t	df	Sig. (2-tailed)	Mean Difference	Std. Error Difference	Lower	Upper
MathScore	Equal variances assumed	.233	.635	-2.283	18	.035	-3.30000	1.44568	-7.46131	.86131
	Equal variances not assumed			-2.283	17.916	.035	-3.30000	1.44568	-7.46359	.86359

a. Message = Genetics

The tables show that using our Bonferroni adjusted alpha of .01, the math test scores of women in the genetics condition ($M = 12.20$, $SD = 3.12$) do not differ significantly from the math test scores of men in the genetics condition ($M = 15.50$, $SD = 3.34$), $t(18) = -2.28$, $p = .03$, $d = 1.02$, 99% CI [$-7.46, 0.86$].

The effect size for this difference is computed as follows:

$$SD_{Pooled} = \sqrt{\frac{(10-1)3.1198^2 + (10-1)3.3417^2}{10+10-2}} = \sqrt{\frac{(9)9.7332 + (9)11.1670}{18}} = \sqrt{10.4501} = 3.2327$$

Using the pooled standard deviation and the means reported in the 'Group Statistics' table, we can now solve for Cohen's d.

$$d = \frac{|15.5000 - 12.2000|}{3.2327} = 1.02$$

Although the effect is not statistically significant (using our Bonferroni adjusted alpha of .01), this is a large effect and indicates that the math test scores of men in the genetics condition are 1.02 standard deviation units higher than those of women in the genetics condition.

Finally, we are ready to consider the effect of gender in the socialization condition. The results are provided in the output window under the heading 'Message = Socialization.'

Message = Socializatioin

Group Statistics[a]

	Gender	N	Mean	Std. Deviation	Std. Error Mean
MathScore	Female	10	16.4000	3.06232	.96839
	Male	10	15.0000	3.05505	.96609

a. Message = Socializatioin

Independent Samples Test[a]

		Levene's Test for Equality of Variances		t-test for Equality of Means					99% Confidence Interval of the Difference	
		F	Sig.	t	df	Sig. (2-tailed)	Mean Difference	Std. Error Difference	Lower	Upper
MathScore	Equal variances assumed	.857	.367	1.023	18	.320	1.40000	1.36789	-2.53738	5.33738
	Equal variances not assumed			1.023	18.000	.320	1.40000	1.36789	-2.53738	5.33738

a. Message = Socializatioin

The tables show that the math test scores of women in the socialization condition ($M = 16.40$, $SD = 3.06$) do not differ significantly from the math test scores of men in the socialization condition ($M = 15.00$, $SD = 3.06$), $t(18) = 1.02$, $p = .32$, $d = 0.46$, 99% CI [−2.54, 5.34].

The effect size for this difference is computed as follows:

$$SD_{\text{Pooled}} = \sqrt{\frac{(10-1)3.0623^2 + (10-1)3.0551^2}{10+10-2}} = \sqrt{\frac{(9)9.3777 + (9)9.3336}{18}} = \sqrt{9.3557} = 3.0587$$

Using the pooled standard deviation and the means reported in the 'Group Statistics' table, we can now solve for Cohen's d.

$$d = \frac{|16.4000 - 15.0000|}{3.0587} = 0.46$$

The Cohen's d value of 0.46 indicates that there is a medium-sized effect of gender in the socialization condition. Specifically, it indicates that the math test scores of women in the socialization condition are 0.46 standard deviation units higher than the math test scores of men in the socialization condition.

Reporting the Results

Taken together, the results of the 2 × 2 between groups factorial ANOVA and the follow-up simple main effects analyses would be reported in the following manner:

A 2 × 2 ANOVA with the message participants overheard (genetics, socialization) and gender (male, female) as between-subjects factors revealed no significant main effect of message, $F(1, 36) = 3.46$, $p = .07$, $\eta^2 = .08$, $\eta_p^2 = .09$, and no significant main effect of gender, $F(1, 36) = 0.91$, $p = .35$, $\eta^2 = .02$, $\eta_p^2 = .02$, on math test scores. However, there was a significant interaction between gender and message, $F(1, 36) = 5.58$, $p = .02$, $\eta^2 = .12$, $\eta_p^2 = .13$.

Simple main effects analyses using a series of independent-samples t tests with a Bonferroni adjusted alpha of .01 revealed that the math test scores of women in the genetics condition ($M = 12.20$, $SD = 3.12$) were significantly lower than the math test scores of women in the socialization condition ($M = 16.40$, $SD = 3.06$), $t(18) = −3.04$, $p = .007$, $d = 1.36$, 99% CI [−8.18, −0.22]. In contrast, the math test scores of men in the genetics condition ($M = 15.50$, $SD = 3.34$) did not differ significantly from the math test scores of men in the socialization condition ($M = 15.00$, $SD = 3.06$), $t(18) = 0.35$, $p = .73$, $d = 0.16$, 99% CI [−3.62, 4.62]. Further, it was discovered that the math test scores of women in the genetics condition did not differ significantly from the math test scores of men in the genetics condition, $t(18) = −2.28$, $p = .03$, $d = 1.02$, 99% CI [−7.46, 0.86]. Similarly, the math test scores of women in the socialization condition did not differ significantly from the math test scores of men in the socialization condition, $t(18) = 1.02$, $p = .32$, $d = 0.46$, 99% CI [−2.54, 5.34].

2 × 3 BETWEEN GROUPS FACTORIAL ANOVA

Assume that after completing the study described above, you decide that you could improve upon it even further by including a control condition in which participants overhear a message unrelated to math abilities. Prior to collecting any data you decide to set alpha at .05. You invite 10 women and 10 men to your lab and have them complete a control condition in which they "accidently overhear" a staged conversation between two professors about their plans for the weekend. Participants are then asked to complete a math test containing 20 questions.

You add these data to the data you collected previously from men and women who completed the genetics and socialization conditions, creating the dataset shown below. You hypothesize that women's math test performance will be affected by the message they overhear, while men's math test performance will be unaffected by these messages.

For this experiment, you have two independent variables (message and gender). One independent variable—message—has 3 levels (genetics, socialization, control), and the other independent variable—gender—has 2 levels (male, female). In addition, the different conditions contain different people, so this would be considered a 2 × 3 between groups factorial design. Accordingly, the data will need to be analyzed using a 2 × 3 between groups factorial ANOVA.

Assume you record the following data:

ID	Message	Gender	Math-Score	ID	Message	Gender	Math-Score	ID	Message	Gender	Math-Score
1	1	0	9	21	1	1	17	41	3	0	10
2	1	0	12	22	1	1	19	42	3	0	11
3	1	0	14	23	1	1	12	43	3	0	12
4	1	0	16	24	1	1	20	44	3	0	17
5	1	0	11	25	1	1	18	45	3	0	11
6	1	0	12	26	1	1	17	46	3	0	13
7	1	0	14	27	1	1	10	47	3	0	15
8	1	0	7	28	1	1	14	48	3	0	10
9	1	0	10	29	1	1	16	49	3	0	9
10	1	0	17	30	1	1	12	50	3	0	11
11	2	0	19	31	2	1	11	51	3	1	9
12	2	0	14	32	2	1	14	52	3	1	15
13	2	0	12	33	2	1	15	53	3	1	17
14	2	0	20	34	2	1	15	54	3	1	11
15	2	0	18	35	2	1	20	55	3	1	12
16	2	0	18	36	2	1	11	56	3	1	15
17	2	0	14	37	2	1	14	57	3	1	12
18	2	0	12	38	2	1	15	58	3	1	18
19	2	0	19	39	2	1	15	59	3	1	19
20	2	0	18	40	2	1	20	60	3	1	16

The data displayed in the first 8 columns (the data for messages 1 and 2) are identical to those used in the preceding example. So, using the dataset you created previously, simply add the data from the third control condition (these data are displayed in the last 4 columns of the table shown above). Go to **Variable View** and use the **Values** column to label message **3** as **Control**. Make sure that the split file function is set to analyze all cases. Once the data are entered, your Data View window should look like the one displayed on the following page.

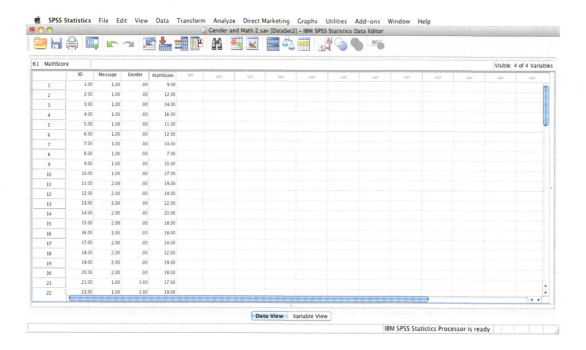

Conducting a Between Groups Factorial ANOVA
(Analyze→General Linear Model→Univariate)

Once again, to conduct the between groups factorial ANOVA, you will need to use the upper toolbar to go to **Analyze→General Linear Model→Univariate**.

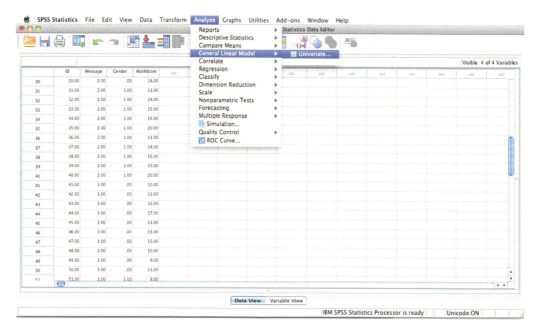

Next a 'Univariate' dialog window like the one shown on the following page will open. Highlight the dependent variable—**MathScore**—by clicking on the variable name and then click on the corresponding **blue arrow** to move it over to the **Dependent Variable box**. Next, highlight the independent variables—**Message** and **Gender**—by clicking on the variable names and move them over to the **Fixed Factor(s) box** by clicking the corresponding **blue arrow**. Next, click on the **Options button**.

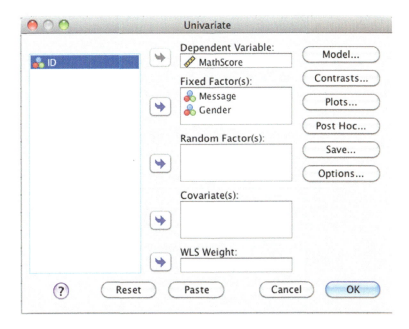

The 'Univariate: Options' dialog window will now appear. Check the options for **Descriptive statistics**, **Estimates of effect size**, and **Homogeneity tests**. Close the dialog window by clicking **Continue**. Finally, click **OK** to close the 'Univariate' dialog window and execute the analysis.

Interpreting the Results

Descriptive Statistics

An output window will now appear displaying a series of tables. Once again, the 'Descriptive Statistics' table shown below displays the means, standard deviations, and sample sizes for females and males in each condition, as well as for the combined groups of males and females. The bottom portion of the table displays descriptive statistics for all of the females (combined across the three conditions), all of the males (combined across the three conditions), and all of the participants (the combined groups of males and females in all three conditions).

Descriptive Statistics

Dependent Variable: MathScore

Message	Gender	Mean	Std. Deviation	N
Genetics	Female	12.2000	3.11983	10
	Male	15.5000	3.34166	10
	Total	13.8500	3.57292	20
Socialization	Female	16.4000	3.06232	10
	Male	15.0000	3.05505	10
	Total	15.7000	3.06251	20
Control	Female	11.9000	2.46982	10
	Male	14.4000	3.27278	10
	Total	13.1500	3.09966	20
Total	Female	13.5000	3.49137	30
	Male	14.9667	3.14570	30
	Total	14.2333	3.37672	60

Levene's Test of Homogeneity of Variance

The table labeled 'Levene's Test of Equality of Error Variances' shown below displays the results of the test of the assumption of homogeneity of variance. The *p* value shown in the table is .71, which is greater than .05, so the assumption of homogeneity of variance has been met.

Levene's Test of Equality of Error Variances[a]

Dependent Variable: MathScore

F	df1	df2	Sig.
.584	5	54	.712

Tests the null hypothesis that the error variance of the dependent variable is equal across groups.

a. Design: Intercept + Message + Gender + Message * Gender

Main Effects and Interaction

The primary results of the analysis are displayed in the table labeled 'Tests of Between-Subjects Effects' shown below.

Tests of Between-Subjects Effects

Dependent Variable: MathScore

Source	Type III Sum of Squares	df	Mean Square	F	Sig.	Partial Eta Squared
Corrected Model	164.933[a]	5	32.987	3.508	.008	.245
Intercept	12155.267	1	12155.267	1292.604	.000	.960
Message	69.433	2	34.717	3.692	.031	.120
Gender	32.267	1	32.267	3.431	.069	.060
Message * Gender	63.233	2	31.617	3.362	.042	.111
Error	507.800	54	9.404			
Total	12828.000	60				
Corrected Total	672.733	59				

a. R Squared = .245 (Adjusted R Squared = .175)

By reviewing the table shown above, you should be able to see that there is a significant main effect of message, $F(2, 54) = 3.69$, $p = .03$, $\eta^2 = .10$, $\eta_p^2 = .12$, while the main effect of gender, $F(1, 54) = 3.43$, $p = .07$, $\eta^2 = .05$, $\eta_p^2 = .06$, is not significant. Moreover, there is a significant interaction between message and gender, $F(2, 54) = 3.36$, $p = .04$, $\eta^2 = .09$, $\eta_p^2 = .11$. This interaction suggests that that the influence of the message participants overheard on their math test scores varies with gender and/or that the effect of gender on math test scores differs depending on the message participants overheard.

Calculating the Effect Sizes (Eta-Squared)

Remember the eta-squared values reported in the preceding paragraph need to be hand calculated by dividing the between groups sum of squares by the total sum of squares (displayed in the row labeled 'Corrected Total').

The eta-squared value for the independent variable message is:

$$\eta^2 = \frac{SS_B}{SS_{Total}} = \frac{69.4333}{672.7333} = .10$$

The eta-squared value for the independent variable gender is:

$$\eta^2 = \frac{SS_B}{SS_{Total}} = \frac{32.2667}{672.7333} = .05$$

The eta-squared value for the interaction is:

$$\eta^2 = \frac{SS_B}{SS_{Total}} = \frac{63.2333}{672.7333} = .09$$

Recall that eta-squared values of .01 are considered small, values of .06 are considered medium, and values of .14 are considered large. Thus, the values reported above indicate that there are medium-sized effects of message and gender on math test scores, and that there is a medium-sized effect of the interaction between message and gender on math test scores.

Simple Main Effects

Since there is a significant interaction between the independent variables, we will need to conduct simple main effects analyses to understand the nature of the interaction. Once again, these analyses involve examining the effect of each independent variable at each level of the other independent variable.

The simple main effects analysis for a 2 × 3 ANOVA can be conducted by performing a series of one-way ANOVAs and independent-samples t tests.[32] Specifically, we will use:

1. A one-way ANOVA to examine the influence of the three messages on women
2. A one-way ANOVA to examine the influence of the three messages on men
3. An independent-samples t test to examine the influence of gender in the genetics condition
4. An independent-samples t test to examine the influence of gender in the socialization condition
5. An independent-samples t test to examine the influence of gender in the control condition

Since we are going to use five separate tests, we should use a Bonferroni correction to alpha in order to reduce our family-wise Type I error rate. If we want to keep our overall alpha level at .05, and we are running five tests, we need to divide our alpha of .05 by 5 (.05/5 = .01) and use the Bonferroni adjusted alpha of .01 to determine statistical significance.

Conducting a Simple Main Effects Analysis for a 2 × 3 Between Groups ANOVA

The Influence of Message (Separated by Gender)

The main effects analysis indicated that the message participants overheard had a significant effect on the combined groups of men and women, but the presence of an interaction once again suggests that the effect of the message may vary with gender. Therefore, we will begin our simple main effects analysis by

---

[32] Once again, we will have to use this quick and dirty method for our simple main effects analysis since the limited student version of SPSS does not contain the necessary syntax editor required to use the more conventional method.

examining the influence of the three messages on women and men separately. Since we want to consider these two groups separately, we will need to use the split file function to split our sample into males and females. Go to **Data→Split File**. Using the 'Split File' dialog window, click **Organize output by groups**, and then move the variable **Gender** into the **Groups Based on box**. Click **OK**.

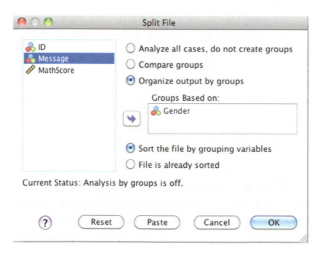

Now that we have split the file into males and females, we can use one-way ANOVAs to separately examine (1) the influence of the three messages on women's math test scores, and (2) the influence of the three messages on men's math test scores. Note that we are using one-way ANOVA because we are comparing the means of three conditions.

As described in Chapter 8, to perform a one-way ANOVA, you need to go to **Analyze→General Linear Model→Univariate**. This will open the 'Univariate' dialog window. MathScore should still appear in the box labeled Dependent Variable, and Gender and Message should still appear in the box labeled Fixed Factor(s). Since we want to examine only the influence of message, you will need to **remove** the variable **Gender** from the **Fixed Factor(s) box** using the corresponding **blue arrow**. The dialog window should now look like the one shown below.

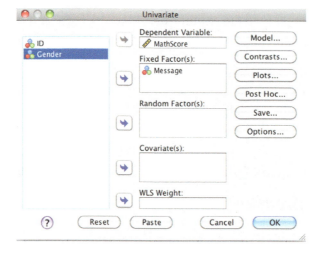

Next, click on the **Options button** to open the 'Univariate Options' dialog window. Check the boxes labeled **Descriptive statistics, Estimates of effect size, and Homogeneity tests**. Using the box labeled '**Significance Level**,' located at the bottom of the dialog window change the .05 default to our Bonferroni adjusted alpha level of **.01**. Click **Continue**.

Finally, click on the **Post Hoc button** in the 'Univariate' dialog window. Move the independent variable **Message** into the box labeled **Post Hoc Tests for**. Check the option for **Tukey**. Click **Continue** and then click **OK** to execute the analysis.

Interpreting the Results

The results of the one-way ANOVAs will now appear in the output window separated by gender. Let's start by considering the results for the women shown under the heading 'Gender = Female.'

Tests of Between-Subjects Effects[a]

Dependent Variable: MathScore

Source	Type III Sum of Squares	df	Mean Square	F	Sig.	Partial Eta Squared
Corrected Model	126.600[b]	2	63.300	7.532	.003	.358
Intercept	5467.500	1	5467.500	650.606	.000	.960
Message	126.600	2	63.300	7.532	.003	.358
Error	226.900	27	8.404			
Total	5821.000	30				
Corrected Total	353.500	29				

a. Gender = Female

b. R Squared = .358 (Adjusted R Squared = .311)

As described completely in Chapter 8, the primary results of interest are displayed in the 'Tests of Between-Subjects Effects' table shown above. The results presented in the row labeled 'Message' indicate that there is a significant main effect of message on the women, $F(2, 27) = 7.53$, $p = .003$, $\eta^2 = .36$. Recall that for one-way ANOVA, the value of eta-squared is identical to the value of partial eta-squared, so we can report just the partial eta-squared statistic presented in the table as an eta-squared statistic.

The presence of a significant main effect of message indicates that at least two of the three means differ from one another. We will now need to examine the results of the post hoc comparisons to determine which means differ significantly. The results pertaining to the post hoc analysis are shown below in the table labeled 'Multiple Comparisons.'

Multiple Comparisons[a]

Dependent Variable: MathScore

Tukey HSD

(I) Message	(J) Message	Mean Difference (I–J)	Std. Error	Sig.	99% Confidence Interval Lower Bound	Upper Bound
Genetics	Socialization	−4.2000[*]	1.29643	.009	−8.3205	−.0795
	Control	.3000	1.29643	.971	−3.8205	4.4205
Socialization	Genetics	4.2000[*]	1.29643	.009	.0795	8.3205
	Control	4.5000[*]	1.29643	.005	.3795	8.6205
Control	Genetics	−.3000	1.29643	.971	−4.4205	3.8205
	Socialization	−4.5000[*]	1.29643	.005	−8.6205	−.3795

Based on observed means.
The error term is Mean Square(Error) = 8.404.

[*]. The mean difference is significant at the

a. Gender = Female

As described in Chapter 8, the results displayed in the 'Multiple Comparisons' table indicate that the math test scores of women in the genetics and socialization conditions differ significantly ($p = .009$), the math test scores of women in the genetics and control conditions do not differ significantly ($p = .97$), and the math test scores of women in the socialization and control conditions differ significantly ($p = .005$).

We will now consider the results of the one-way ANOVA for the men shown under the heading 'Gender = Male.' Once again, the 'Tests of Between-Subjects Effects' table shown below contains the primary results of interest.

Tests of Between-Subjects Effects[a]

Dependent Variable: MathScore

Source	Type III Sum of Squares	df	Mean Square	F	Sig.	Partial Eta Squared
Corrected Model	6.067[b]	2	3.033	.292	.749	.021
Intercept	6720.033	1	6720.033	645.927	.000	.960
Message	6.067	2	3.033	.292	.749	.021
Error	280.900	27	10.404			
Total	7007.000	30				
Corrected Total	286.967	29				

a. Gender = Male

b. R Squared = .021 (Adjusted R Squared = −.051)

As displayed in the row labeled 'Message,' there is not a significant main effect of message for the men, $F(2, 27) = 0.29$, $p = .75$, $\eta^2 = .02$. Now we know that the interaction indicates that the message participants overheard had a significant effect on women's math test scores but not on men's math test scores.

The absence of a significant main effect of message for men tells us that none of the means differ significantly. As such, it is not necessary (nor appropriate) to conduct a post hoc analysis. Therefore, we will simply disregard the 'Multiple Comparisons' table with the results of the post hoc analysis for males.

The Influence of Gender (Separated by Message)

The main effects analysis indicated that the overall math test performance of men and women did not differ. However, the presence of an interaction once again suggests that the effect of gender on math test scores may differ depending on the message participants overheard. To explore this possibility, we will need to examine the influence of gender in the genetics, socialization, and control conditions separately. Since we want to consider these three conditions separately, we will need to use the split file function to split our sample into the three message conditions. Go to **Data→Split File**. Using the 'Split File' dialog window, move the variable **Gender** *out* of the **Groups Based on box**, and move the variable **Message** *into* the **Groups Based on box**. Click **OK**.

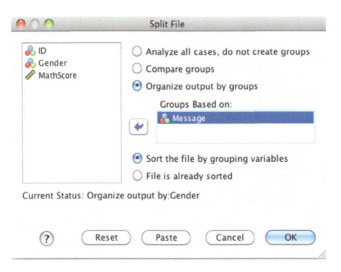

Now that we have split the file into the three message conditions, we can use independent-samples *t* tests to compare the women's and men's math test performance in each condition separately. To perform this series of analyses, go to **Analyze→Compare Means→Independent-Samples T Test**. Move the dependent variable **MathScore** into the **Test Variable(s) box**, and the independent variable **Gender** into the **Grouping Variable box**. Click **Define Groups**. Since we coded the genders with 0s and 1s, we will need to enter **0** in the **Group 1 box** and **1** in the **Group 2 box**. Click **Continue** to close the dialog window.

Finally, since we are using a Bonferroni adjusted alpha of .01 to determine statistical significance, we will need to change the confidence interval to the corresponding 99% confidence interval. Click on the **Options button** in the 'Independent-Samples T Test' dialog window. Using the 'Independent-Samples T Test: Options' dialog window, change the 95% confidence interval default setting to 99% by typing **99** into the **Confidence Interval Percentage box**. Click **Continue** and then **OK** to close both dialog windows and execute the analysis.

Interpreting the Results

The results of the three *t* tests will now appear in the output window. Let's start by considering the results for the genetics condition shown under the heading 'Message = Genetics.'

Message = Genetics

Group Statistics[a]

	Gender	N	Mean	Std. Deviation	Std. Error Mean
MathScore	Female	10	12.2000	3.11983	.98658
	Male	10	15.5000	3.34166	1.05672

a. Message = Genetics

Independent Samples Test[a]

		Levene's Test for Equality of Variances		t-test for Equality of Means					99% Confidence Interval of the Difference	
		F	Sig.	t	df	Sig. (2-tailed)	Mean Difference	Std. Error Difference	Lower	Upper
MathScore	Equal variances assumed	.233	.635	-2.283	18	.035	-3.30000	1.44568	-7.46131	.86131
	Equal variances not assumed			-2.283	17.916	.035	-3.30000	1.44568	-7.46359	.86359

a. Message = Genetics

The tables show that, using our Bonferroni adjusted alpha of .01, the math test scores of women in the genetics condition ($M = 12.20$, $SD = 3.12$) do not differ significantly from the math test scores of men in the genetics condition ($M = 15.50$, $SD = 3.34$), $t(18) = -2.28$, $p = .03$, $d = 1.02$, 99% CI [-7.46, 0.86]. Remember that the Cohen's *d* value needs to be hand calculated. You should practice calculating this value on your own.

Next, we will consider the results shown under the heading 'Message = Socialization.'

Message = Socialization

Group Statistics[a]

	Gender	N	Mean	Std. Deviation	Std. Error Mean
MathScore	Female	10	16.4000	3.06232	.96839
	Male	10	15.0000	3.05505	.96609

a. Message = Socialization

Independent Samples Test[a]

		Levene's Test for Equality of Variances		t-test for Equality of Means					99% Confidence Interval of the Difference	
		F	Sig.	t	df	Sig. (2-tailed)	Mean Difference	Std. Error Difference	Lower	Upper
MathScore	Equal variances assumed	.857	.367	1.023	18	.320	1.40000	1.36789	-2.53738	5.33738
	Equal variances not assumed			1.023	18.000	.320	1.40000	1.36789	-2.53738	5.33738

a. Message = Socialization

The tables show that the math test scores of women in the socialization condition ($M = 16.40$, $SD = 3.06$) do not differ significantly from the math test scores of men in the socialization condition ($M = 15.00$, $SD = 3.06$), $t(18) = 1.02$, $p = .32$, $d = 0.46$, 99% CI [-2.54, 5.34]. You should practice calculating the Cohen's *d* value on your own.

Finally, we will consider the results of the control condition shown under the heading 'Message = Control.'

Message = Control

Group Statistics[a]

	Gender	N	Mean	Std. Deviation	Std. Error Mean
MathScore	Female	10	11.9000	2.46982	.78102
	Male	10	14.4000	3.27278	1.03494

a. Message = Control

Independent Samples Test[a]

		Levene's Test for Equality of Variances		t-test for Equality of Means						
									99% Confidence Interval of the Difference	
		F	Sig.	t	df	Sig. (2-tailed)	Mean Difference	Std. Error Difference	Lower	Upper
MathScore	Equal variances assumed	1.513	.235	-1.928	18	.070	-2.50000	1.29658	-6.23212	1.23212
	Equal variances not assumed			-1.928	16.741	.071	-2.50000	1.29658	-6.26499	1.26499

a. Message = Control

The tables show that the math test scores of women in the control condition ($M = 11.90$, $SD = 2.47$) do not differ significantly from the math test scores of men in the control condition ($M = 14.40$, $SD = 3.27$), $t(18) = -1.93$, $p = .07$, $d = 0.86$, 99% CI [−6.23, 1.23]. Once again, you should practice calculating the Cohen's d value on your own.

Reporting the Results

These results could be reported in the following manner:

A 2 × 3 ANOVA with gender (male, female) and message participants overheard (genetics, socialization, control) as between-subjects factors revealed a significant main effect of the message participants overheard, $F(2, 54) = 3.69$, $p = .03$, $\eta^2 = .10$, $\eta_p^2 = .12$, but no significant main effect of gender, $F(1, 54) = 3.43$, $p = .07$, $\eta^2 = .05$, $\eta_p^2 = .06$, on math test scores. Of greater interest, there was a significant interaction between gender and the message participants overheard, $F(2, 54) = 3.36$, $p = .04$, $\eta^2 = .09$, $\eta_p^2 = .11$.

Simple main effects analyses using a series of independent-samples t tests with a Bonferroni adjusted alpha of .01 revealed that the math test scores of women in the genetics condition ($M = 12.20$, $SD = 3.12$) did not differ significantly from the math test scores of men in the genetics condition ($M = 15.50$, $SD = 3.34$), $t(18) = -2.28$, $p = .03$, $d = 1.02$, 99% CI [−7.46, 0.86]. The math test scores of women in the socialization condition ($M = 16.40$, $SD = 3.06$) also did not differ significantly from the math test scores of men in the socialization condition ($M = 15.00$, $SD = 3.06$), $t(18) = 1.02$, $p = .32$, $d = 0.46$, 99% CI [−2.54, 5.34]. Similarly, the math test scores of women in the control condition ($M = 11.90$, $SD = 2.47$) did not differ significantly from the math test scores of men in the control condition ($M = 14.40$, $SD = 3.27$), $t(18) = -1.93$, $p = .07$, $d = 0.86$, 99% CI [−6.23, 1.23].

As predicted, additional simple main effects analyses revealed no significant main effect of message for the men, $F(2, 27) = 0.29$, $p = .75$, $\eta^2 = .02$. In contrast, there was a significant main effect of message for the women, $F(2, 27) = 7.53$, $p = .003$, $\eta^2 = .36$. Post hoc analyses using Tukey's *HSD* test revealed that the math test scores of women in the genetics condition were significantly lower than the math test scores of women in the socialization condition ($p = .009$). The math test scores of women in the control condition were also significantly lower than the math test scores of women in the socialization condition ($p = .005$). Finally, there was no significant difference in the math test scores of women in the genetics and control conditions ($p = .97$).